THE LIFE OF DAVID LACK

The Life of David Lack

FATHER OF EVOLUTIONARY ECOLOGY

Ted R. Anderson

OXFORD
UNIVERSITY PRESS

Oxford University Press is a department of the University of Oxford.
It furthers the University's objective of excellence in research,
scholarship, and education by publishing worldwide.

Oxford   New York
Auckland   Cape Town   Dar es Salaam   Hong Kong   Karachi
Kuala Lumpur   Madrid   Melbourne   Mexico City   Nairobi
New Delhi   Shanghai   Taipei   Toronto

With offices in
Argentina   Austria   Brazil   Chile   Czech Republic   France   Greece
Guatemala   Hungary   Italy   Japan   Poland   Portugal   Singapore
South Korea   Switzerland   Thailand   Turkey   Ukraine   Vietnam

Oxford is a registered trade mark of Oxford University Press
in the UK and certain other countries.

Published in the United States of America by
Oxford University Press
198 Madison Avenue, New York, NY 10016

Library of Congress Cataloging-in-Publication Data
Anderson, Ted R.
The life of David Lack : father of evolutionary ecology / Ted R. Anderson.
   pages cm
Summary: "Most people who have taken a biology course in the past 50 years are familiar with the work of
David Lack, but few remember his name. Almost all general biology texts produced during that period have a figure showing
the beak size differences among the finches of the Galapagos Islands from Lack's 1947 classic, Darwin's Finches. Lack's pioneering
conclusions in Darwin's Finches mark the beginning of a new scientific discipline, evolutionary ecology. Tim Birkhead, in his
acclaimed book, The Wisdom of Birds, calls Lack the 'hero of modern ornithology.' Who was this influential, yet relatively
unknown man? The Life of David Lack, Father of Evolutionary Ecology provides an answer to that question based on Ted
Anderson's personal interviews with colleagues, family members and former students as well as material in the extensive
Lack Archive at Oxford University"—Provided by publisher.
Summary: "This book examines the life of scientist David Lack"—Provided by publisher.
ISBN 978-0-19-992264-2 (hardback)
1. Lack, David, 1910–1973. 2. Ornithologists—Great Britain—Biography. 3. Biologists—Great Britain—Biography.
4. Birds—Evolution. 5. Evolution (Biology) I. Title.
QL31.L27A53   2013
598.092—dc23
[B]
2012046180

9 8 7 6 5 4 3 2

Printed in the United States of America
on acid-free paper

Contents

Preface

⟨ornament⟩ ———

DAVID LACK (1910–1973) is arguably the father of evolutionary ecology. With the publication of both *Darwin's Finches* (Cambridge University Press) and "The Significance of Clutch-Size" (*Ibis* 89:302–352) in 1947, Lack inaugurated the application of the Modern Synthesis (the successful integration of Darwinian natural selection with Mendelian genetics that was forged in the 1930s and early 1940s) to the field of ecology. With the publication of these seminal works, as well as numerous subsequent books and papers, this British ornithologist must be acclaimed as at least first among equals of the major progenitors of this subdiscipline of ecology that has commanded so much attention for the past half-century. Although his name is known to every ornithologist, his role in the development of evolutionary ecology is less appreciated in the broader ecological community than those of G. Evelyn Hutchinson, Robert MacArthur Gordon Orians, and other pioneers of the subdiscipline, particularly in North America. It seems appropriate, at this time marking the centenary of Lack's birth, to attempt to rectify this deficiency with this biography that focuses on his life and work.

A few years ago, I was discussing the extensive work of Peter and Rosemary Grant on the geospizine finches of the Galapagos Islands with a colleague who has also worked extensively on the islands. I stated that I had read much of their earlier work but had not kept up on some of their later work, in part because I had come to the conclusion that their findings were "just an extended exposition of David Lack" (here, I am sure that I owe an apology to the Grants). My colleague replied without hesitation, however: "Isn't that what we all do?" Although this is undoubtedly an overstatement that ignores major theoretical advances since Lack, it nevertheless illustrates a significant point: Much of contemporary evolutionary ecology derives in many ways from the pioneering work of Lack. It is no doubt for this reason that Tim Birkhead described Lack as the "hero" of 20th century ornithology in *The Wisdom of Birds* (Bloomsbury, 2008).

Another dimension of the story is the remarkable collection of evolutionary biologists and ecologists that were present in Oxford during the nearly 28-year period that Lack served as Director of the Edward Grey Institute of Field Ornithology. In addition to Lack, this group included Arthur J. Cain, Dennis Chitty, Charles Elton, E. B. Ford, Alister Hardy, Niko Tinbergen, and George C. Varley. Together these brilliant mentors and their many students contributed dramatically to the reshaping of biology. Although some were more theoretically oriented, the work of all was firmly grounded in the study of animals in the field.

One aspect of Lack's life that many today might consider anachronistic is that, although he was one of the leading evolutionary biologists of his time, he converted to Christianity at the age of 38. He remained an active Anglican communicant for the rest of his life, and he wrote and spoke publicly about the unresolved conflicts between evolutionary biology and Christianity. Indeed, his most important contribution on the subject, *Evolutionary Theory and Christian Belief, the Unresolved Conflict*, has recently been reprinted by Taylor & Francis (2008) in their Philosophy of Science series.

David Lack wrote 13 books, ranging from major scientific treatises to books aimed at the general reader. I have organized this biographical sketch into 13 chapters titled with the names of his books in their chronological order. Each chapter describes the major scientific contributions of the book for which it is titled, and Lack's life story is woven into the chapters in a roughly chronological manner, although some of the chapters are topical rather than chronological.

Acknowledgments

⌒——

I WISH TO thank first and foremost the Lack family, all of whom have cooperated wholeheartedly in seeing this project through to completion. Elizabeth Lack invited us into the Lack home on Boars Hill and granted two lengthy interviews. Each of the four Lack children was interviewed either in person or by telephone and was most helpful and encouraging. The two older sons, Peter and Andrew, also assisted by providing leads and contact information for other sources of information on their father, and Andrew was able to elicit some answers to questions I posed to David's sister Katreen before her death. Both also read and commented on drafts of some chapters of the book, particularly those dealing directly with the family, and they saved me from numerous potentially confusing or embarrassing errors of fact (particularly on English geography, which is not taught in American public schools).

Special appreciation is also due to the Edward Grey Institute of Field Ornithology and its director, Prof. Ben Sheldon, who provided me access to the David Lack Archive housed in the Alexander Library of Ornithology in Oxford (references to materials housed in the archive will refer to it as the Lack archive). Sophie Wilcox, Subject Librarian and Manager of the Alexander Library, was also extremely helpful, as were other of the Alexander's librarians. The text includes numerous references to scientific papers written by David Lack that identify only the journal in which each was published and the year of publication. A complete bibliography of Lack's scientific publications is available from the Lack archive at the Alexander Library.

I also am eternally grateful to the many students, colleagues, neighbors and friends of David Lack who provided much of the material for this biography. I conducted personal interviews, either in person or by telephone, with the following individuals: R. J. Andrew, N. P. Ashmole, H. V. Bergamini, W. R. P. Bourne, the late D. Chitty, the late J. R. Clarke, L. Cole, A. W. Diamond, J. B. Falls, M. P. L Fogden, R. Gillmor, R. A. Hinde, P. Honeychurch, L. Houlden, D. Jenkins, A. K. Kepler, C. Kepler, C. J. Krebs, Lord J. R. Krebs, G. Lambrick, the late R. Leacock, E. Leopold, B. MacArthur, J. MacArthur, J. F. Monk, W. W. Murdoch, J. B. Nelson, I. Newton, D. Nichols, R. Obermier, G. H. Orians, R. Overall, S. (Church) Parker, C. M. Perrins, M. Perrins, Lord N. Rea, U. N. Safriel, D. A. Scott, N. Slack, B. (Macpherson) Sladen, W. J. L. Sladen, the late D. W. Snow, B. Stonehouse, D. Summers-Smith, Lady K. (Russell) Tait, P. Thomas, E. (Kabraji) Todds, M. M. (Betts) Turner, the late M. E. Varley, H. (Russell) Ward, A. Watson, and T. Williams. Many quotations attributed to these individuals either by name or by their relationship to David Lack (e.g., son, neighbor, D.Phil. student) are not accompanied by endnotes.

The archivists of several institutions were also very helpful: L. Larby, Gresham's School, Holt; R. Hyam, Archivist, and P. Grimstone, Sub-Librarian, Magdalene College, Cambridge; Y. Widger, Dartington Hall Trust; C. Hopkins, Trinity College, Oxford; and R. Darwall-Smith, Magdalen College, Oxford. I also thank T. Leech for his gracious hospitality during my brief visit to Gresham's School.

Several individuals kindly read sections of the manuscript at various points and offered suggestions and made important corrections. I am deeply appreciative of the comments made by the four Lack children, W. Richter, A. K. Kepler, and C. M. Anderson. Any errors of fact or awkwardness in presentation remain, however, the sole responsibility of the author.

Finally, I thank the individuals at Oxford University Press who have greatly facilitated the publication of *The Life of David Lack*. These include Jeremy Lewis, Science Editor; Hallie Stebbins, Assistant Editor; Marc Schneider, Production Editor; and Beverly Braunlich, Copyeditor.

THE LIFE OF DAVID LACK

There are frequent false starts before the final departure.
—*The Birds of Cambridgeshire*, p 34

1

The Birds of Cambridgeshire

⌔──

THE CAMBRIDGE BIRD Club published David Lack's first book, *The Birds of Cambridgeshire*, a year after he completed his studies at Magdalene College, Cambridge. Although it provides few hints as to the ultimate influence this British ornithologist was to have on 20th century ecology, it does contain some glimpses into a mind that was the product of a changing scientific landscape in the field of ornithology, and indeed in biology as a whole. *The Birds of Cambridgeshire* is not simply a compilation of the species that had been recorded in the county. Instead, Lack devoted almost half of the book to chapters describing the major habitats in the county, observations on migration, historical changes in the county's avifauna, and the resident status of the 160 regularly occurring species. He expressly de-emphasized rare and accidental species, a significant departure from the typical emphases of ornithologists at the time.

Kristin Johnson[1] has written about a fundamental transformation in what constituted accepted scientific ornithology during the period 1920–1950, as reflected in the articles published in *Ibis*, the journal of the British Ornithologists' Union. Johnson described this as a transformation from museum-based geographical ornithology to university-based population ecology; it involved a shift from specimen-based research in geographical variation, leading to a proliferation of sub-specific nomenclature, to field studies of the behavior and ecology of the living bird. In the process, the criteria for what constituted legitimate scientific study of birds were transformed. David Lack was a pivotal figure in this transformation, but in 1934 the revolution was in its infancy.

Who was this man who exerted such a profound influence during his lifetime and whose ideas continue to generate both discussion and controversy? He was quintessentially English and spent most of his professional life as Director of the Edward Grey Institute of Field Ornithology at Oxford. He is arguably the father of evolutionary ecology, certainly one of its central figures, and yet, as his widow Elizabeth said, "He was a definite Christian."[2] This chapter deals with his formative years, including his family life and formal schooling through university.

FAMILY

David Lack was born in London on July 16, 1910, the first of four children of Dr. Harry Lambert Lack and his wife, Kathleen. Lambert Lack, as he was generally known, was one of the most distinguished rhinolaryngologists of his time. Born in 1867, he had grown up on his father's farm in Norfolk but was inspired by his uncle, a physician in the county, to pursue a career in medicine. He was educated at King's College, London (M.R.C.S., L.R.C.P., 1890), and at the University of London (M.B., 1892), and he was elected Fellow of the Royal College of Surgeons in 1893. He became a pioneer in his field, initiating new surgical techniques and publishing a classic book in laryngology, *The Diseases of the Nose and its Accessory Sinuses* (Longmans Green, 1906). His name is attached to a metal tongue depressor with a spoon-shaped end that he developed, the "Lack." In 1908 he married Kathleen Rind.

Kathleen was the daughter of Col. McNeill Rind, formerly of the Indian Army, and was of Scotch-Irish descent. She was in the legitimate theater, primarily in traveling companies, but ended her acting career when she married. She had been an active supporter of women's suffrage and believed that women should have a profession; so, although she gave up the stage, she continued to teach elocution and verse reading. She and Lambert shared literary interests, and they often hosted meetings of the Poetry Society in their home. Kathleen's sister, Ada, a professional singer who never married, also lived with the Lacks. Kathleen's niece, Gillie, who visited Kathleen with her mother late in Kathleen's life, remembered her as "a friendly woman, perhaps on the serious side . . . not much sense of humour." Ada, on the other hand, "was much more flamboyant and colourful."[3] Besides David, the Lack's children were Kathleen (affectionately called Katreen by family and friends alike), born in 1912; Christofer, born in 1914; and Oliver, born in 1918.

The Lacks lived in Harley Street at the time of David's birth, but a year later they moved to a spacious house in Devonshire Place. The house had a dumbwaiter to carry food from the kitchen in the basement to the nursery on the top floor. The Lacks employed seven servants and a chauffeur. David remembered that primary

FIGURE 1: Kathleen Lack with her four children: Christofer in the foreground, Oliver on Kathleen's lap, Katreen standing on left, David on right. (Photograph courtesy of Lack family.)

care of the four children was performed "by a succession of nannies, each of whom left when we had become emotionally dependent on her."[4] This statement reflects the fact that David believed that he had been abandoned by his mother during his childhood, when she devoted only a half-hour each evening to being with the children. This perception may account for his attempt to be a much more engaged parent for his own children, one of whom reported that "he never really forgave his mother."[5]

Summer holidays were spent at a large 17th-century house in New Romney on the Kent coast. The house had a 4-acre garden in which Kathleen organized an annual family production of the fairy scenes from *A Midsummer Night's Dream*. David played the role of Bottom. Katreen remembered that she played Titania and that a friend of Kathleen's, the Shakespearean actress, Beatrice Wilson, often came down from London to participate in the production. Beatrice performed regularly in the annual Shakespeare Birthday Festival presentation of excerpts from the bard's plays at the Old Vic Theatre in London. Katreen also recalled that the New Romney holiday was the one time of the year when the busy Lambert Lack was usually present and able to be with his children.[6]

It was also in the spacious garden at New Romney that David learned to identify many of the common birds. In later years, as his passion for birds deepened, he spent many hours exploring the nearby Romney Marsh, a 100-square-mile marsh with a diverse avifauna and a checkered past. Formed as a delta of the River Rother, its proximity to the continent meant that it had historically served as both an invasion route and a haven for smugglers. (Although portions of the marsh had been utilized for centuries for various forms of agriculture, intensive agriculture after World War II reduced the marsh to a fraction of its original size, with only fragments preserved today.)

David was schooled at home until the age of 7 years, when he was enrolled at the Open Air School in Regent's Park. In September 1920, he was enrolled in an acclaimed day school, The Hall in Hampstead, which he attended until the age of 13. His final year of preparatory school was spent at Foster's School, a boarding school in the Hampshire village of Stubbington (now a part of Fareham). This first separation from his family was an unhappy experience, and after a miserable year at Foster's, during which he "vowed to become a kindly schoolmaster,"[7] he was enrolled at Gresham's School in Holt near the Norfolk coast.

The trip from London to Holt in September 1924 was via the London and North Eastern Railway from Liverpool Street Station through Norwich and Sherringham. Today, one can get a glimpse of the experience by traveling the final 4 miles from Sherringham to Holt aboard the "Poppy Line" steam train, operated by North Norfolk Railway. If one arrives at the end of April, as I did in 2008, the admonition that "the only thing between the Norfolk Coast and Siberia is the North Sea" may become apparent. As the train from Norwich to Sherringham approaches the Norfolk coast, the broad plains clothed in a mosaic of verdant green wheat fields and brilliant yellow fields of rape in bloom morph into wooded coastal hills, the temperature drops, and scattered lilacs, hesitating as they begin to bloom, are the only hints of the coming spring. The Norfolk coast yields reluctantly to the inevitable advance of spring. The market town of Holt is within easy walking distance of the Holt station, and between the station and the town, hidden from view by an extensive deciduous woodland, lies the modern Gresham's School.

GRESHAM'S SCHOOL

The English public school has often both mystified and captivated Americans. Part of the mystification is the fact that "public school," like "suspenders," has a different meaning in America than in England. As Shaw said, in America we haven't spoken English for years. The English public school is essentially equivalent to an elite private boarding school in America, such as Choate Academy or Phillips Academy Andover, and has no resemblance to American public schools.

Sir John Gresham, who was born in Holt, Norfolk, founded the school bearing his name in 1555. Gresham was a successful London merchant who became Lord Mayor of London in 1547. He established an endowment of land and investments for Gresham's School, which he entrusted to the Worshipful Company of Fishmongers, a London guild; the guild has exercised administrative responsibility for the school since Gresham's death in 1556. The school survived the fire that destroyed much of Holt on May Day in 1708, but it remained small and relatively obscure until it was expanded dramatically under the leadership of George W. S. Howson, its innovative headmaster from 1900 until his death in 1919.

Howson not only enlarged the physical facilities of Gresham's but also placed his individual stamp on the school's educational character and environment. He chose to de-emphasize the classics (the curricular mainstay at most public schools at the time), instead emphasizing science and also providing more latitude for the study of music and art. Sports were given a much lower priority at Gresham's than at other contemporary schools, being entirely intramural with no matches played against other schools. Participation in Officers' Training Corps was not compulsory, as it was at most other public schools.

Another staple of English public schools, hazing by older boys, was strongly discouraged, and corporal punishment was almost nonexistent. In its place, Howson introduced an Honour Code that required all boys to pledge to their housemasters:

(1) not to swear, smoke, or say or do anything indecent; (2) to report their own transgressions to the housemaster; and (3) to report transgressions of other boys if they failed to self-report. After Howson's sudden death in January 1919, the deputy headmaster, J. R. Eccles, was appointed headmaster, a position he held until 1935. Eccles, a scientist by training, continued and extended Howson's policies.

Boys were encouraged to pursue hobbies, and Gresham's had a wide range of clubs to support these interests. In Lack's schooldays, these included the Debating Society, Literature Society, Society of Arts, Sociological Society, Camera Club, and even the League of Nations Union. By far the most popular club, however, was the Natural History Society, to which more than two-thirds of the boys belonged. The Natural History Society had sections on botany, entomology, geography, geology, meteorology, ornithology, physics and chemistry, and zoology. It also sponsored field trips during the summer term and frequent lectures from outside speakers. It published an annual report on its activities and sponsored the annual Holland-Martin Natural History Prize.

The Gresham's of the mid-1920s was widely regarded as one of the more progressive schools of the time, and it was a favorite choice of those interested in such an environment for their young men. When David Lack arrived in 1924, after the unhappy year at Foster's, Gresham's had an enrollment of approximately 250 boys, most boarding in one of four houses. David was assigned to Woodlands, the house in which headmaster Eccles also served as housemaster. The other houses were Old School House, Farfield, and Howson's (named for the late headmaster). Old School House, the oldest of the houses, was in Holt, whereas the three newer houses were located on an expansive campus on the outskirts of the town.

The student population of Lack's time at Gresham's was remarkable by any standard, probably because of the popularity of the school among the liberal intelligentsia. W. H. Auden, who was beginning his upper sixth form when David arrived, was a resident of Farfield House. He would later write that the *raison-d'etre* of the English public school was "the mass production of gentlemen,"[8] noting that the development of a social consciousness that might threaten the position of the English upper crust was hardly expected to be part of its curriculum.

Other Greshamians during Lack's time who later achieved prominence included Alan Hodgkin, Christopher Cockerell, Benjamin Britten, and Jocelyn Edward Salis (J. E. S.) Simon. Hodgkin, who attended from 1927 to 1932, went on to Trinity College, Cambridge, where he ultimately collaborated with Sir Andrew Fielding Huxley in elucidating the electrophysiological basis of nerve cell transmission. The two shared the Nobel Prize in Physiology or Medicine in 1963 for this achievement. Hodgkin was knighted in 1972 and served as President of the Royal Society from 1970 to 1975. Cockerell (1924–1929) was the son of the curator of the Fitzwilliam

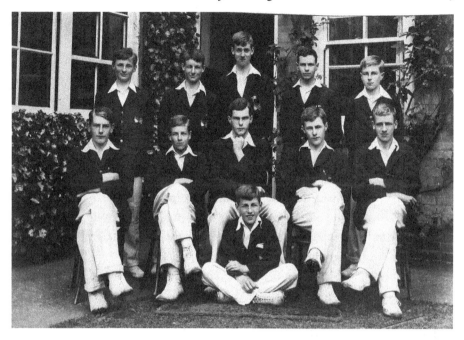

FIGURE 2: Woodlands Second XI cricket team, 1925, with David standing on right. (Photograph courtesy of Lack family.)

Museum, Cambridge. His father despaired of the intellectual future of his son, whose most acute interest, with house guests including such luminaries as George Bernard Shaw, Joseph Conrad, Siegfried Sassoon, and T. E. Lawrence, was in Lawrence's motorcycle. Cockerell studied engineering at Peterhouse, the oldest of the Cambridge colleges (founded in 1284), and became a pioneering engineer in early radio and television communications and radar; he is most famous as the inventor of the hovercraft. He was elected a Fellow of the Royal Society in 1967 and was knighted in 1969.

Although Benjamin Britten received a music scholarship to attend, the music teacher at Gresham's apparently remarked sarcastically "So *you* are the little boy who loves Stravinsky!" upon meeting him, and later informed him that he had no future in music.[9] Little wonder, therefore, that Britten spent only 2 years at Gresham's before accepting a scholarship to the Royal College of Music in 1930. He became one of the major composers of the 20th century, and, after earlier turning down a knighthood, was knighted in 1976, shortly before his death. J. E. S. Simon (1920–1928) studied law at Trinity Hall, Cambridge, and went on to distinguished careers in both politics and the judiciary, serving as a conservative Member of Parliament and a minister in the Churchill and MacMillan governments and rising to the Presidency of the Probate, Divorce, and Admiralty Division of the High Court in 1962. He joined the House of Lords in 1972 as Lord Simon of Glaisdale, and he died in 2006.

Another contemporary, and with Lack a resident of Woodlands, was the son and namesake of the Liberal Member of Parliament, Sir Donald Maclean, who sent three of his sons to Gresham's. The younger Donald excelled both academically and athletically, and he and David were teammates on the Woodlands rugby team in the fall term of 1928. Donald was prefect of Woodlands during his final year before receiving a scholarship to Trinity Hall, Cambridge.

One of the friendships Maclean established at Gresham's was with James Klugman, the son of a wealthy Jewish rope manufacturer in London. As a schoolboy, the younger Klugman was an ardent Communist, a fact which, along with his Jewishness, meant that Donald thought he would be a less than welcome guest in the Maclean home. Donald did, however, spend some of his school holidays with James at his home in Hampstead. James went on to study at Trinity College, Cambridge, and was active in the growing Communist cell there. After earning a First in French and German at Cambridge, he spent most of the rest of his life working for various Communist organizations and wrote the two-volume *History of the Communist Party of Great Britain* (Lawrence & Wishart, 1968, 1969). The extent of Klugman's influence on Maclean's ultimate career path is not known.

In Lack's short autobiographical sketch published in *Ibis* at the time of his death, Lack described Maclean as a "former Foreign Office official,"[10] a droll understatement of the actual reality. Maclean had indeed risen through the ranks of the British Foreign Service, and in May 1951, when he suddenly defected to the Soviet Union with Guy Burgess, he was serving as head of the Foreign Office in Washington, D.C. The facts of the matter, revealed more than a decade later, were that Maclean and Burgess were members of the so-called Cambridge Five, arguably the most effective spy ring of all time. The name of the group derived from its origin at Cambridge in the early 1930s, when Maclean, Burgess, and two of the other members (Kim Philby and John Cairncross) were students and the fifth, Anthony Blunt, was a Fellow of Trinity College. The full extent of the ring's activities in providing information to the Soviets about both British and American military capabilities and plans, spanning nearly a quarter of a century from the beginning of World War II until Philby's defection in 1963, will probably never be known. Maclean died in Moscow in 1983. As we shall see later, Lack's relationship with Maclean at Gresham's was not his only tangential encounter with the Cambridge Five.

Lack later described his time at Gresham's in somewhat unflattering terms, but it is clear that he discovered his vocation while there and that the relatively open atmosphere of the school provided him the opportunity and latitude to pursue his true passion, the study of birds. He mentioned only two teachers by name, W. H. Foy and G. H. Lockett. Foy, from whom Lack received his first biology lessons, "was too eccentric to be allowed to stay, and returned to malaria research."[11] Lockett was chemistry master and an authority on spiders. During one summer term, he convinced Eccles to permit him to take three of the boys on local field trips on one of the three cricket days each week. As Lack recalled, "We used to bicycle in line past the headmaster's window, but as soon as we reached the Holt Lowes, we separated for the afternoon, reassembling to return together."[12] Presumably, this subterfuge was intended to preserve the illusion that the boys had been supervised for the entire afternoon.

Lack said little about his formal studies at Gresham's, stating that he had only worked at about three-quarters capacity. He did win set prizes in biology, maths, and Latin but said that his English was mediocre and his French deplorable. He lamented the fact that he had not studied more mathematics.

Lack's extracurricular activities at Gresham's included rugby, cricket, and chess as well as participation in the Debating Society, the Camera Club, and, most importantly, the Natural History Society. He was in the Officers' Training Corps through the fifth form but was one of four boys who did not participate in the sixth form. Instead, during the sixth form, he manifested his first interest in another lifelong pursuit by singing in the choir. He was a member of the Woodlands rugby and cricket teams for several years, but his level of enthusiasm for participation is

FIGURE 3: Photograph of Woodlands House residents, 1925, with Headmaster J. R. Eccles fifth from left in second row and David second from left seated on the ground. (Photograph courtesy of Lack family.)

apparent in his autobiographical sketch: "Fortunately, I had my appendix out at the start of the rugger term of 1926, and was off games for the term."[13] This happenstance permitted him to devote game days to the study of birds in the environs around the school.

During the first summer term at Gresham's in 1925, Lack participated in Natural History Society–sponsored field trips to Scolt Head and Hickling Broad, excursions that fuelled his nascent interest in birds. Because of this interest, the Natural History Society continued to be at the center of his activities while he was at Gresham's. He participated in almost all of the field trips sponsored by the society, and he served as secretary during his final year there. He spent much of his free time pursuing his passion for birds. Alan Hodgkin, who was also interested in birds, recalled his first meeting with Lack on the salt marshes at Cley, Norfolk, where he had walked on a bird-watching trip one Sunday afternoon: "I could tell that he was from the same school because he was wearing the regulation Sunday uniform of black suit, stiff collar and straw hat."[14] They spent a number of Sunday afternoons thereafter watching birds, and Lack later enlisted Hodgkin to help search for nightjar nests on Kelling Heath during the summer months.

David's detailed chronicles of the activities of the zoology section of the Natural History Society, published in *The Annual Report of the Gresham's School Natural History Society*, identified a number of students in addition to Hodgkin who were active bird watchers. Two were brothers, A. P. (Peter) and M. F. M. (Maury) Meiklejohn, both of whom went on to study at Oriel College, Oxford. Peter became a physician and was senior lecturer in nutrition for 15 years at the University of Edinburgh. Maury became Stevenson Chair of Italian at Glasgow University and during 25 years in that position built the department into the largest in Britain. He continued his intense interest in birds, publishing frequently in *Scottish Naturalist* and *Scottish Birds*, and is also widely remembered for a weekend column in the *Glasgow Herald* (over the initials MFMM). He may be rightly regarded as the first "twitcher" as well as the inventor of the birding Life List.

The Natural History Society sponsored an annual essay competition for original natural history observations, the Holland-Martin Natural History Prize. David won the prize in 1926 with an essay entitled, "Three Birds of Kelling Heath." The essay contained many original observations on three species that bred on Kelling Heath near Holt—the nightjar, the redshank, and the ringed plover. (The original handwritten essay is in the Lack archive at the Edward Grey Institute of Field Ornithology in Oxford). Accompanying the description of each species was a reproduction of the evolutionary tree for birds copied from W. P. Pycraft's *History of Birds* (Methuen, 1910), with a red line added to trace the phylogeny of the family to which the species belonged. The 15-year-old Lack's evolutionary perspective was evident throughout the essay. The three species featured in the essay are all ground-nesting species. Two, the redshank and the ringed plover, have precocial young, whereas the nightjar has altricial young (i.e., hatchlings are helpless and require feeding and intensive care by the parents). In the essay, Lack speculated about the origins of altricial development in birds, assuming that the precocial pattern is the primitive form based on the fact that the reptilian ancestors of birds have precocial hatchlings. He stated,

> Those birds however which remained in the trees developed young helpless at birth. There are many reasons for this. One will do here. Those young which wandered further from the nest when danger came, on the re-arrival of the mother with food, would get less food than those, less active, which had remained nearer; so by a process of natural selection, the weaker survived.[15]

Just the kind of conclusion that would fire the imagination of a schoolboy: In the struggle for existence, the strong do not always prevail!

On his winter holiday in 1925, David began to keep a bird diary, a habit he continued for the next dozen years. The family was spending the Christmas holiday in New Romney, and David's first entry, dated December 24, was unexciting in the extreme: "On the shore in the estuary found the remains of a <u>curlew</u> and 2 <u>guillemots</u>. No live birds save gulls of different species."[16] He routinely underlined species names of birds he observed, presumably to be able to find records more easily. Pollution was already a problem, as indicated by an entry dated January 2, 1926, still from New Romney: "Saw an 'oiled-up' grebe on the shore, a <u>great-crested</u>. It tried to fly but fell on its breast after about 10 yards."

The Royal Society for the Protection of Birds (RSPB) sponsored an annual essay contest for public school boys. Lack won the Silver Medal (first prize in the senior division) in two successive years. In 1927 he won for his essay, "The Birds of Greatstone," about the birds of Romney Marsh, which was based on his observations during the family's holidays in New Romney. The winning essay in 1928 was entitled "The Nightjar" and was based on further observations of this species on Kelling Heath. The latter essay was published in its entirety in *The Annual Report of the Gresham's School Natural History Society*. That same year, his brother Christofer, also a student at Gresham's, won the top prize in the junior division of the RSPB contest. Lambert Lack asked his older son not to enter the contest the following year, saying, "Give someone else a chance to win." Bruce Campbell won the top prize that year. Campbell later became a leading British ornithologist, earning the first doctorate in the country for field work on birds (Edinburgh, 1947) and serving as the first full-time Secretary of the British Trust for Ornithology, a position he held for 10 years beginning in 1948.

The young David failed to comply with his father's wish that he follow his foot-steps into the medical profession, instead deciding to become a zoologist. Because the only professional positions open to zoologists at that time were as entomologists, Dr. Lack decided to provide his son with some practical experience before he went up to the university. He arranged for David to spend the summer term of 1929 working at a museum in Frankfurt am Main. David lived with a family there for 4 months while pinning insects at the Senckenberg Museum, a job he found "ex-tremely" boring. There were three benefits of his time in Germany, however: he learned to speak German "fluently but inaccurately"; he was introduced to orchestral classical music; and, most personally satisfying, he learned his continental birds on weekend field trips.

MAGDALENE COLLEGE, CAMBRIDGE

"Cambridge was a spring awakening after the winter of a public school"[17]—So Lack described the transition from Gresham's to Cambridge. Although he failed to gain a scholarship to the university, he did gain admission to Magdalene College, where he began to read zoology in Michaelmas term, 1929. He saw his first wood sandpiper at the Cambridge sewage farm on the day of his arrival in October, and as the records in *The Birds of Cambridgeshire* attest, the sewage farm became one of his frequent haunts. At the time Lack was a student, Magdalene was home to about 160 undergraduates.

Magdalene College traces its origins to 1428, when the Order of Saint Benedict founded a hostel for monk-scholars on the north bank of the River Cam. Anne Nev-ille, the Duchess of Buckingham, was apparently the college's first benefactor, spon-soring the building of the First Court during the 1470s. (Courts are the Cambridge equivalent of Oxford's quadrangles.) The hostel became known as Buckingham College, a name it retained until the English Reformation. Buckingham College officially ceased to exist when the Benedictines surrendered Crowland Abbey to Henry VIII in 1539, but it was refounded in 1542 by Henry's Lord Chancellor, Thomas Lord Audley of Walden. (Audley had presided over the trial of Sir Thomas More in 1835). Henry VIII could be identified as the founder of Magdalene College, because by his grant the First Court of the former Buckingham College became the First Court of the new college.

Probably the most famous alumnus of Magdalene's half-millennium of history prior to Lack's matriculation was the diarist, Samuel Pepys. Pepys, who graduated from Magdalene in 1654, recorded some of the most turbulent events of England's long history, including the Restoration of the monarchy, the London Plague, and the Great Fire of London. He also described in carefully encrypted language a number of personal peccadilloes. He bequeathed his library of more than 3000 books to Magdalene, and the Pepys Library is still an integral part of the college.

One of Magdalene's claims to fame is the fact that C. S. Lewis was Professor of Medieval and Renaissance Literature at the college from 1954 until his death in 1963. Lewis is perhaps best known as the author of *The Chronicles of Narnia* and as a major Christian apologist of the 20th century. Porters at Magdalene are quick to point out the second floor windows in the east wing of the First Court where Lewis lived in the former Master's Quarters. Magdalene College literature decries the fact that *Shadowlands*, the widely viewed film about Lewis and his wife, Joy Gresham, fails to portray his association with Cambridge, focusing solely on his former connection with Magdalen College, Oxford.

Lack's contemporaries at Magdalene included the actor Michael Redgrave; the mathematician and World War II Special Forces officer at Bletchley Park, Dennis Babbage; and the last English Governor-General of Australia and First Viscount De L'Isle, William Phillip Sydney—all third-year students when Lack matriculated. The future Lord Justice of Appeal, Roualeyn Cumming-Bruce, matriculated in Lack's third year.

One of Lack's chief mentors at Magdalene was the Pepys Librarian, Francis McDougall Charlewood Turner. Turner was also the college organist. The young Lack at this time was an agnostic, but he became a choral exhibitioner: "We had anyway to attend chapel twice weekly," he recalled later.[18] He also sang in college madrigals as his interest in music continued to develop. He received a small choral scholarship during his later years at Magdalene.

His interest in poetry and literature deepened as well. Turner had inaugurated spring reading parties at Mortehoe, a village on the Devonshire coast. Lack described the parties as "exhilarating," and he continued to attend them even after leaving Cambridge. He displayed his wide acquaintance with classical literature and poetry in his later writings, particularly in *The Life of the Robin* and *Robin Redbreast*.

Lack found his formal studies much less exhilarating, however. Zoology at Cambridge in Lack's day consisted primarily of comparative morphology and invertebrate embryology, subjects that he considered irrelevant or at best tangential to his true interest—the evolution, ecology, and behavior of birds. Cambridge has a unique system for awarding baccalaureate degrees, requiring students to complete two parts (I and II) of the Tripos (possibly so-called for the three-legged stool on which students once sat during their oral examinations). Lack read zoology, botany, and geology for part I of the Tripos and zoology for part II. He received Seconds in the oral examinations at the end of each of the parts, earning his B.A. degree in 1933. His part II examiner wrote the following to Lack's Magdalene tutor on June 16, 1933:

Dear Tutor at Magdalene,

You may wish to hear about Lack. He would doubtless have obtained a First Class if he had not more or less collapsed in his Practical. I think his was a case of sensitiveness and worry. Undoubtedly the man is First Class but you know that examiners cannot consider the mental state of an examinee, for it is far too dangerous.

Yours Sincerely
J. Stanley Gardner[19]

Lack was disappointed with this result, but in a letter written 33 years later to another student who had also earned a Second at Cambridge, he added a handwritten postscript: "If it's any encouragement, I also got one class lower than expected at Cambridge. It was a great *help* to my future!"[20]

David returned to Cambridge in January 1935 to pay the £5 fee required to obtain his Cambridge M.A. He was subsequently awarded a Sci.D. from Cambridge in 1949, primarily for his work on the Galapagos finches (see Chapter 3).

One salient feature of the Cambridge of Lack's day, its political activism, failed to capture his interest. As indicated earlier, Trinity College was the center of an active Communist cell. Julian Heward Bell, a student at King's College and the nephew of Virginia Wolff, wrote in the *New Statesman* in 1933:

In the Cambridge that I first knew, in 1929 and 1930, the central subject of ordinary intelligent conversation was poetry. As far as I can remember, we hardly ever thought or talked about politics .... By the end of 1933, we have arrived at a situation in which almost the only subject of conversation is contemporary politics, and in which a very large majority of the intelligent undergraduates are Communists or near Communists.[21]

The Cambridge University Socialist Society, founded in 1905 as the Cambridge University Fabian Society, was also very prominent. The Cambridge University Eugenics Society, which had been founded by R. A. Fisher in 1911 when he was an undergraduate, was resuscitated in 1930 after a hiatus of several years and remained active until 1933. Lack paid little attention despite the fact that his unofficial mentor, Julian Huxley, was deeply involved in the British Eugenics Society.

Instead, the "Cambridge spring" that he described had much more to do with the freedom Cambridge provided him to pursue the passion for birds that he had developed at Gresham's. He read voraciously and was particularly influenced by Julian Huxley's studies of bird behavior and Edmund Selous' *Realities of Bird-life* (Constable,

1927). In "My Life as an Amateur Ornithologist," he also mentioned that the first non-ornithological writings in zoology that interested him were works on evolution by J. B. S. Haldane and R. A. Fisher, two of the first to articulate elements of the emerging Modern (or neo-Darwinian) Synthesis, the synthesis that finally established the central role of natural selection as the primary mechanism of evolutionary change.

But his true passion was pursuing birds in the field. He joined the Cambridge Ornithological Club (which he succeeded in having renamed the Cambridge Bird Club when he was president) and spent innumerable hours exploring the bird life of the region. Other student members of the club during his time in Cambridge included Peter Scott, son of the fabled Antarctic explorer; Arthur Duncan, later a leading Scottish ornithologist; and Tom Harrisson, who pioneered the study of the birds of Sarawak on the island of Borneo. Another member was Dominic Serventy, who was to become an outstanding seabird biologist for whom the Royal Australian Ornithologists' Union named its annual award for distinguished work on birds of the Australasian region.

David returned to Kelling Heath near Holt in June 1930 to continue his observations on nightjars. He witnessed the previously undescribed precopulatory display of the nightjar, which he identified as his most exciting ornithological observation to that date. The description of the display became the subject of his first serious scientific paper, which was published in *Ibis* in 1932.

Although David was not "fired" by the Zoology Department at Cambridge, he did receive inspiration from the Botany Department. In 1931, Dr. Alexander Stuart Watt, a lecturer in forestry who was coordinating a study of plant communities in state-funded afforestation projects in East Anglia, invited David to study changes in the bird community with time since planting. The results of this research prompted David to question how birds choose their appropriate habitat and led him to write an important paper on habitat selection in birds (*Journal of Animal Ecology*, 1933).

CAMBRIDGE EXPEDITIONS

David participated in Cambridge-sponsored expeditions during the summers of 1931, 1932, and 1933. The first was a joint Oxford-Cambridge expedition to St. Kilda, an isolated, windswept archipelago located 40 miles west of the Outer Hebrides in the North Atlantic. St. Kilda had been inhabited since at least the late Middle Ages, but the confluence of a number of circumstances had led to the request for evacuation of the remaining 36 St. Kildans the previous year. *The Times* of August 29, 1930, stated, "The evacuation affects some 36 natives, together with the island mules and the missionary and his small family." The 1931 expedition involved six scientists and six returning St. Kildans, including the island's last postmaster, Alec Ferguson. The St. Kildans were retrieving belongings abandoned in the evacuation. David's responsibility was to collect insects. He wrote a short piece for the February 1932 issue of *Magdalene College Magazine*, entitled "St. Kilda Sabbath," in which he described the Sunday services in the St. Kilda church. Ferguson led successive Gaelic and English services attended by all, and one of the other St. Kildans, Finlay Macqueen, uttered long, soulful prayers in Gaelic. Lack co-authored two scientific papers based on the expedition's findings, one on early fall bird migration in *Scottish Naturalist* (1932) with John Buchan and T. H. Harrisson, and the other with Harrisson on the breeding birds of the island, also in *Scottish Naturalist* (1933).

The following year, David invited Colin Bertram, an undergraduate at St. John's College, Cambridge, to accompany him on a two-person expedition to the Arctic. Bertram, who was also reading zoology, would later be a member of the British Graham Land Expedition to the Antarctic Peninsula (1934–1937) and would serve as Director of the Scott Polar Research Institute from 1949 to 1956. In June 1932, the two young explorers sailed on the steam trawler *Arkwright* of Hull, and on June 18 they crossed the Arctic Circle en route to Bear Island. David wrote an account of the expedition for the February 1933 issue of *Magdalene College Magazine*. Four Norwegian meteorologists were the sole inhabitants of the island, which lies midway between the northern tip of Norway and the island of Spitsbergen. Although Bear Island receives more than 100 days of continuous sunlight, David reported that only 5 of the 52 days spent on the island were clear, with the remainder blanketed by a freezing fog. He described most of the island as a barren, boulder-strewn plateau with the tallest vegetation, Iceland poppies, being only 6 inches high. The Norwegians frequently cooked for the two zoologists, serving meals of fresh cod, seabirds (mostly guillemots), and whale beef (obtained from whalers working the coast). David described these meals as a "refreshing addition to mostly tinned provisions."[22] A Polish expedition, which included a Dutchman and a French Swiss leader in addition to several Poles, arrived near the end of their stay on the island. A final supper on the island, which included the Norwegians and the members of the two expeditions, ended with the five nationalities represented singing their respective national anthems. The conversation was primarily in German, which was not the first language of any of the participants but which most could understand. Bertram co-authored two papers with David based on their Bear Island observations, one in *Ibis* (1933), on the island's birds, and the other in *Geographical Journal* (1933).

In the summer of 1933, after concluding his Cambridge studies, Lack helped organize a Cambridge-sponsored expedition to Iceland and East Greenland. His primary interest was in exploring nunataks, the exposed rocky ridges bordering glaciers or ice fields, to identify what animals might have survived the Pleistocene

FIGURE 4: Finlay MacQueen on the Sabbath, St. Kilda, 1931. (Photograph by David Lack, courtesy of Lack family.)

FIGURE 5: Colin Bertram (left) and David on Bear Island, 1932. (Photograph courtesy of Lack family.)

glaciations. He, Bertram, and Brian Roberts comprised the three-person expedition that sailed aboard the *Pourquoi Pas?* IV (French for "Why not?"). The ship, a three-masted barque designed for polar exploration, was built for the famous French scientist and polar explorer, Jean-Baptiste Charcot, who was heading the current expedition. Charcot was known as the "Polar Gentleman."

The *Pourquoi Pas?* arrived at Reykjavik, Iceland, on July 8, and while it was refueling and waiting for the ice to go out in Scoresby Sound (East Greenland), the three Cambridgemen headed north across Iceland, eventually spending 5 days on Grímsey. This island lies on the Arctic Circle, 35 miles off the north coast of Iceland, and was inhabited at the time by about 130 people. In a paper describing their observations on Icelandic birds, Lack and Roberts stated that the inhabitants of Grímsey "subsist mainly by fishing and bird-snaring," eating "adult Guillemots and Puffins and young Fulmars, Gannets, and Kittiwakes."[23]

After resuming its journey, the *Pourquoi Pas?* reached Scoresby Sound on July 28, and a few days later the trio set up a base camp on Hurry Inlet, a fjord running 40 miles northward from the Sound. For the next two and a half weeks, they made numerous sorties from the base camp into Jameson Land and Liverpool Land (across the inlet from their camp). They reported their bird observations in a paper in *Ibis* (1934), which included a discussion of non-breeding in several species and its possible relationship to the low point in the lemming cycle (no lemmings were observed during their entire stay). They expressed appreciation in the paper for Charles Elton's contributions to the discussion of non-breeding species. Thus began the long and complicated relationship between David and Charles (see Chapter 11).

The expedition returned to England aboard the *Pourquoi Pas?* in September, and David took up his post as biology master at Dartington Hall School (see Chapter 2). In an undated letter addressed to "My dear Roberts, Bertram and Lack," Charcot wrote:

> Shakespeare, if my memory is good, wrote amongst many other fine things that what he hated most in the world was ingratitude. I partake of this hatred. Such a reproach cannot go to all three of you and I am extremely sensible for your very touching demonstration; the more so that in fact having you in the old ship's staff was a great pleasure and I only hope that it may be renewed in still better conditions. Let me tell you very simply and sincerely what I say frequently that all three of you are the nicest young fellows alive, that I wish you to always consider me as a good old friend and that on my side I know that I can count on you. Do not forget that my home on land or water is equally yours.[24]

FIGURE 6: Sketch of David Lack preparing to go into the field in East Greenland, 1933. Sketch made by either the Polar Gentleman (Jean-Baptiste Charcot) or the captain of the *Pourquoi Pas?* (Courtesy of Lack family.)

The reference to "better conditions" presumably referred to weather conditions experienced by the three members of the expedition in East Greenland, which are hinted at by a drawing of Lack made there by either Charcot or the ship's captain. Tragically, Charcot was to lose his life 3 years later when the *Pourquoi Pas?* ran aground south of Iceland while returning from another Greenland expedition. There was only one survivor among the 40 men aboard.

[W]hen scientists abolish the gods of earth, of lightning and of love, they create instead gravity, electricity and instinct. Deification is replaced by reification, which is only a little less dangerous and far less picturesque.

—*The Life of the Robin*, p 168

2

The Life of the Robin

THUS DAVID LACK introduced his discussion of instinct in his charming second book, *The Life of the Robin*. The robin is probably Britain's best known, and undoubtedly its most beloved, bird species; in fact, it was voted Britain's national bird in 1960. The book is based on 4 years of field study of the species at Dartington Hall in south Devon, where Lack was teaching during the 1930s. He completed the manuscript in 1942, and the book was published in 1943 by H. F. & G. Witherby Ltd. of London. In *The Life of the Robin*, Lack combined scientific description and analysis with literary and historical references to the subjects discussed, a style that makes the book accessible to the nonprofessional without sacrificing its scientific rigor. One evidence of the attempt to strike this balance is the use of the terms "cock" and "hen" instead of "male" and "female," which he attributed to a suggestion from Annabel Williams-Ellis (author of *The Arabian Nights* and *All Stracheys are Cousins*, among more than 50 books). He met Annabel, who was born Mary Annabel Nassau Strachey, when he was teaching at Dartington Hall School. She was the mother of one of his students there.

The Life of the Robin went through five editions between 1943 and 1970, including a Pelican paperback edition in 1953 and a Fontana New Naturalist paperback edition in 1970. The fourth edition, published in hard cover by Witherby in 1966, was illustrated by Robert Gillmor. *The Life of the Robin* has had a wide readership and is undoubtedly Lack's most widely read book outside the professional ornithological community. One of his sons says that he is still often asked by nonprofessionals whether he is related to the Lack who wrote *The Life of the Robin*.

Lack inaugurated intensive field studies of the robin on a 20-acre study area on the grounds of Dartington Hall in January 1935, a little more than a year after he arrived at Dartington Hall School as science mentor. He trapped adult robins and color-ringed them with unique combinations for individual recognition. One of his students who assisted in capturing and ringing the robins recalled, "I remember one Robin caught on May 25th 1935 and nicknamed 'Jubilee' and with red, white and blue rings. That detail was too flippant for the book!"[1] Lack also built two outdoor aviaries to study the behavior and breeding biology of pairs of captive robins, making the first observation of captive breeding in the species. He also used stuffed robins, and various parts of stuffed robins, placed in robin territories to observe the behavior of the territory-holding birds. A Dartington Hall film student, Richard Leacock, took movies of the responses of the resident robins.

Lack's study paralleled an earlier study of the robin in Ireland by J. P. Burkitt, and Lack made frequent comparisons of his results with Burkitt's. He also made numerous comparisons of his results with those of Margaret Morse Nice in her classic study of the song sparrow in North America. Allusions to the works of Oskar Heinroth, Konrad Lorenz, and Niko Tinbergen show that he was conversant with the pioneering work in ethology being done on the continent in the 1930s.

Topics that Lack discussed in detail included the function of song, pair formation and breeding biology, displays, the function of territory, migration, and population dynamics. He had already published scientific papers covering some of this material in *Proceedings of the Zoological Society of London* (1939), *Ibis* (1940), and *British Birds* (1940). In *The Life of the Robin*, he attempted to present a comprehensive description of the ecology and behavior of the species by addressing the complete annual cycle in south Devon. Some of his conclusions presage his later positions, developed more fully and extended in other works.

Song is one of the most salient features of robin life, and from his observations of color-ringed birds of known sex, Lack discovered that both males and females sing during autumn and winter. At those times, both sexes defend individual territories against all other individuals. In late winter, when pair formation begins, males increase the frequency and intensity of their singing, but females curtail their singing. Males continue to sing after they are mated, although the frequency and intensity of song decrease. Lack concluded that the primary function of song is in territorial defense, although it also plays a role in mate attraction for unmated males.

Lack's findings with regard to migration in the robin were quite startling. He found that adult females are more likely to migrate than adult males, and that almost all juveniles migrate. With regard to the role of territory in robin life, Lack concluded that territory does not control the size of the breeding population. He continued to maintain this position throughout his life, and it became one of the points of sharp disagreement during his later debates with V. C. Wynne-Edwards about group selection (see Chapter 9).

One of the most controversial of Lack's conclusions in *The Life of the Robin* concerned the robin's mortality. Based on his observations of ringed robins, he concluded that 72 percent of juveniles and 62 percent of adults alive on August 1 do not survive until the next August 1. These estimates of mortality rates in a small bird were much higher than most ornithologists of the day believed them to be, and many disagreed sharply with his conclusions. Some nonprofessionals hearing his findings for the first time expressed their disapproval pugnaciously. A story that was still circulating around the Bureau of Animal Population in Oxford 20 years later described one such encounter after a presentation by Lack to a local bird club. An elderly woman approached him and insisted that the same robin had been living in her back garden for 17 years. When Lack tried patiently to explain that ringing would be required to establish this observation with certainty and that it was highly unlikely that it had been the same robin in her garden for that length of time, she began to beat him on the head with her umbrella. Subsequent work on many other bird species showed that he was correct, and the implications of such a high mortality rate for the efficacy of natural selection to act on genetic variation in populations were strongly supported.

In a very positive review of *The Life of the Robin*, H. N. "Mick" Southern of the Bureau of Animal Population in Oxford was perhaps the first to recognize Lack's skill in communicating scientific material to a general audience. He wrote, "[T]he style of writing is straightforward and readily understandable to non-scientific readers without being reduced to baldness."[2]

The remainder of this chapter deals with the period from 1933 to October 1945, when David assumed the directorship of the Edward Grey Institute of Field Ornithology in Oxford. It covers the salient events during his years as a schoolmaster at Dartington Hall School and his service during World War II. As an exception, the discussion of Lack's Galapagos expedition of 1938–1939, will be deferred until Chapter 3.

DARTINGTON HALL SCHOOL

In March 1933, William Burnlee "Bill" Curry, Headmaster of Dartington Hall School, wrote to V. S. Vernon Jones inquiring about the suitability of David Lack to serve as science mentor at Dartington Hall. Julian Huxley had learned of the school's need for a science mentor and had suggested that the position be filled by an ecologist; he also suggested David as an appropriate candidate. The Magdalene College archive does not contain a copy of Jones' response, but it was presumably positive, because Lack took up the post for the fall term that same year. In his autobiographical sketch, "My Life as an Amateur Ornithologist," Lack said that during his unhappy year at Foster's School he had resolved to become a kindly schoolmaster, a resolution that he repudiated while at Gresham's, when he decided to become a zoologist instead of a teacher. Dartington Hall School provided him the opportunity to fulfill the earlier resolve.

Dartington Hall is located on the River Dart near Totnes, 5 miles from Devon's eastern coast. The Estate dates to at least the year 833, when it was listed in the registry of Shaftesbury Abbey as the Homestead on the Meadow by the River Dart. After the Norman Conquest, it was given to one of William the Conqueror's companions-in-arms, William de Falaise. Other owners followed, but the estate finally became property of the Crown for want of an heir.

Richard II presented the Estate at Dartington to his half-brother, John Holland, in 1384. Holland was a chief advisor to Richard and also a renowned jouster, the latter fact relevant to his plans for the construction of a great manor house on the Estate. Construction began in 1386 and was completed 14 years later. The house was laid out as a double-quadrangle with two interior courts (one large enough to serve as a jousting yard). Only one of the courts remains today. Henry IV deposed Richard in 1399, and John Holland was later beheaded for participating in a conspiracy against Henry. The Holland family eventually lost possession of the estate during the Wars of the Roses.

In 1534 the Estate was acquired by Sir Arthur Chamernowne, whose descendants retained ownership for almost 400 years. What remained of the Estate was purchased in 1925 by Leonard and Dorothy Elmhurst.

The Elmhursts were an unusual pair. Leonard was the second of eight sons of a parson and landowner in west Yorkshire with family roots in the region dating back 600 years. After reading history and theology at Cambridge, he was posted to India during World War I. There he became acquainted with an American missionary, Sam Higginbottom, founder of the Allahabad Agricultural Institute. Higginbottom was devoted to improving the lives of the rural poor by teaching improved agricultural practices. As a result of his wartime Indian experiences, Leonard decided to devote his life to the improvement of agricultural practices, both in England and abroad. After working his way across the Atlantic as a ship's waiter, he studied agriculture at Cornell University's College of Agriculture, receiving a degree in 1921. He returned to India at the invitation of Rabindranath Tagore, founder of the International University at Shantiniketan in West Bengal and 1913 Nobel Laureate in Literature, whom he had met at Cornell. Leonard started the Department of Rural Reconstruction at the university but left India in 1924, once the department was established. During his studies at Cornell, he had also met Dorothy Whitney Straight, a recent widow and mother of three children.

Dorothy was born in Washington, D.C., the daughter of American financier and statesman, William Collins Whitney, and his first wife, Flora. Her mother died when she was just 6 years old, and after her father died when she was 17, she inherited a fortune. She developed a strong social consciousness, working in settlement houses and welfare organizations and serving as the first president of the National Junior League, attempting, as she said, "to keep alive in the privileged members of society the sense of responsibility for those less fortunate than themselves."[3] She also worked in the women's suffrage movement and traveled widely. On a trip to Peking, she met an American banking representative and diplomat, Willard Straight. They married in Geneva, Switzerland, in 1911 and had two sons and a daughter between 1912 and 1916. They also founded the politically liberal magazine, *The New Republic*, in 1914. Willard died in Paris in 1918 from complications of the Spanish flu while helping to make arrangements for the arrival of the American representatives to the Paris Peace Conference. His will requested that something be done to make Cornell University, from which he had graduated in 1901, a more humane place. It was while she was trying to fulfill this request, by building the Willard Straight Student Union at Cornell, that Dorothy met Leonard Elmhurst.

Leonard and Dorothy married in April 1925. They purchased Dartington Hall later that year with the objective of establishing a model for rejuvenation of the rural economy that would halt the flight of people from rural areas in England. The Estate consisted of 820 acres–two farms totaling 600 acres, about 190 acres of woodlands, and the buildings and grounds of the Hall. Their primary aims for the Estate were to establish model programs of agriculture, horticulture, and forestry, including derivative industries such as textiles, sawmilling, and cider making. By 1929, when Dartington Hall Ltd. was incorporated, the Estate was operating two farms, orchards, a poultry industry, a textile mill, and a saw mill.

A secondary aim for the Estate was the establishment of a boarding school, a project whose genesis was based in part on the need to educate Dorothy Elmhurst's three young children: Whitney Willard Straight, born in 1912; Beatrice Whitney Straight, born in 1914; and Michael Whitney Straight, born in 1916. The children had been attending a progressive, experimental school in New York, and Dorothy, whose educational views had been shaped in part by the American educational philosopher, John Dewey, wished to provide a similar schooling opportunity for them in England. Dartington Hall School was founded as a progressive, coeducational boarding school with nine students (including Dorothy's three children) in September 1926. It ceased operations in 1987. From the beginning, the school served the children of workers on the Estate as well as boarding students from other parts of England and a few from abroad.

Leonard prepared two documents describing the goals for the school, *Outline of an Educational Experiment* and *Prospectus*. An overarching principle was articulated in the *Outline*: "For us it is vital that education be considered as life, and not merely as a preparation for life."[4] Four subsidiary principles guided the early development of the school: (1) the curriculum should flow from children's own interests; (2) learning by doing; (3) adults should be friends, not authority figures; and (4) the school should be a self-governing commonwealth. In practice, the Dartington Hall environment was the antithesis of the traditional public school of the day; there was

no corporal punishment, indeed no punishment at all; no prefects; no uniforms; no Officers' Training Corps; no segregation of the sexes; no compulsory games, no compulsory religion, or compulsory anything else; no more Latin, no more Greek; no competition; no jingoism.[5]

Michael Young, an early student, described the school as "a magnet for devotees of naked bodies and cabbage juice."[6] The widespread local belief that the negatives included "no clothes" was a greatly exaggerated misperception, but Bill Curry would later joke, about a man who said that he had been greeted at the door of Dartington Hall by a naked butler: "How did he know it was the butler?" The Elmhursts envisioned the school as a large extended family, and one of their beliefs was that children should have no concerns about seeing each other naked on a daily basis—bathrooms and showers were therefore coeducational.

The school grew slowly, having 25 students in 1929 when Michael Young matriculated. Young maintained a lifelong association with the school, including serving for many years on its Board of Trustees. Young first went into politics, including penning the *Labour Manifesto* which articulated the Labour Party vision that helped reshape England under Clement Attlee's Labour government (1945–1951). Later he wrote an account of the Elmhursts and their grand vision for Dartington Hall (*The Elmhursts of Dartington: The Creation of an Utopian Community*, Routledge & Kegan Paul, 1982). Young was still a student in 1931 when the first headmaster of the school, Bill Curry, was appointed, and he chronicled some of the ensuing changes under Curry's leadership.

Curry came to Dartington Hall from Oak Lane Country Day School, a progressive school in Philadelphia. Born and educated in England, Curry had been recruited to serve as headmaster of Oak Lane after teaching physics for a short time in 1922 at Gresham's School in Holt and serving as senior science master at Bedales School in rural Hampshire. Leonard Elmhurst would later write:

Gifted with a brilliant mind, Bill Curry was a confirmed liberal of the old school, deeply concerned over the way in which ideas of liberty, freedom and democracy might find lively expression in the daily routine of school life.[7]

Curry was a strong-minded and articulate man whose personal and educational philosophies melded readily with those of the Elmhursts. He was an ardent pacifist and an unabashed advocate of progressive education. His commitment to pacifism helped stimulate the establishment of a strong pacifist group at Dartington Hall and also found expression in his active support of the Federal Union Movement, which developed in the years just before the advent of World War II. He believed that education, which was usually sponsored or supported by the state, was generally used "for the deliberate inoculation of nationalism."[8] Curry dated his interest in progressive education to his reading of Bertrand Russell's *Principles of Social Reconstruction* (Allen and Unwin, 1917) in 1919, and it was a great source of satisfaction for him when Russell enrolled his children, John and Kate, in Dartington Hall School in the mid-1930s. He later stated his own educational principles as follows:

> These then were the two intertwined strands of thought out of which my philosophy of education arose: a profound concern that the child be treated with respect and love and never merely as a means to other people's ends, something to be moulded; and an almost desperate concern for peace, and a kindlier and friendlier world, which made me passionately unwilling to be responsible for perpetuating in any way the forms of education which seemed to make the task of creating a peaceful world more difficult.[9]

One change that Curry brought to Dartington Hall School, based in part on his earlier educational posts, was a commitment to upgrading the school's curriculum so that graduates would be competitive for university admissions. He served as headmaster until his retirement in 1967.

One salient feature of Dartington Hall Estate was the Sunday Evening Meeting, which was held in one of the large halls on most Sunday evenings. The meetings brought together students from the school, workers on the estate, and interested neighbors and were devoted to concerts, lectures, debates, recitals, and other intellectual and cultural events. During the period 1930–1935, for instance, lecturers at the Sunday Evening Meetings included Bertrand Russell ("The Revolt against Reason" and "The Case for Complete Pacifism"), Aldous Huxley ("Religion in the Modern World"), A. S. Neill (headmaster of Summerhill, another progressive school), Sir Norman Angell (winner of the 1933 Nobel Peace Prize), H. G. Wells, George Bernard Shaw, and Julian Huxley. It may have been on the occasion of his lecture, on birds, that Huxley became aware of the opening for a science mentor at Dartington Hall School.

The Straight children proved to be three of the more distinguished alumni of Dartington Hall School. Willard was a Grand Prix race car driver, an R.A.F. pilot during World War II, and, later, chief executive officer of British Overseas Airways Corporation and deputy chairman of Rolls-Royce. He was also a Member of Parliament. He died in 1979. Beatrice became a well-known actress of stage and screen in the United States, winning a Tony Award for her portrayal of Elizabeth Proctor in the original Broadway production of Arthur Miller's *The Crucible* in 1953 and an Oscar as Best Supporting Actress in *Network* in 1976. She died in 2001.

Michael Straight's journey after leaving Dartington Hall School was particularly convoluted. He read economics at Trinity College, Cambridge, and was active in Trinity's Socialist Society. In 1935 he visited the Soviet Union and met the young Slade Professor of Fine Arts at Trinity, Anthony Blunt, who was also visiting at the time. Blunt, a member of the Cambridge Five along with Kim Philby, Guy Burgess, Donald Maclean, and John Cairncross, recruited Michael to spy for the Soviet Union. His spying career was brief and inconsequential, except that in 1963, when he was approached about taking a senior arts position in the Kennedy administration, he went to the U.S. FBI and eventually to MI-5 with his knowledge of Blunt's espionage activities. Blunt was eventually identified publicly as a member of the Cambridge Five and was stripped of his knighthood, which he had received in 1956 for serving as Surveyor of the Queen's Pictures.

Michael returned to the United States in 1937 and worked as a speech writer for President Franklin Roosevelt and then as editor of the *New Republic*. He served in the U.S. Army Air Force during World War II. After the war, he was publisher of the *New Republic* until 1956, when he resigned to write full-time. He later served as Deputy Chairman of the National Endowment of the Arts from 1969 to 1977. He died in 2004.

Other notable Dartington Hall alumni from Lack's time at the school include Lucian Freud, Eva Ibbotson, and Richard Leacock. Freud, the grandson of Sigmund, was one of three children of the architect, Ernst Ludwig Freud, who matriculated at Dartington Hall in 1933, after their parents became some of the first to flee Hitler's Germany. Lucien was one of Britain's most preeminent and best-known painters of the last half-century. Eva Ibbotson, born Maria Wiesner, also fled Germany with her father, the physiologist Berthold Wiesner. She became a well-known writer of children's books. Leacock, whose father owned banana plantations in the Canary Islands, became one of the first students in the fledgling film program at Dartington Hall and went on to become a pioneering documentary filmmaker in the United States. As noted earlier, he also filmed some of Lack's work on the robin, and he later accompanied Lack to the Galapagos (see Chapter 3).

SCIENCE MENTOR AT DARTINGTON HALL

Following his expedition to Iceland and Greenland in the summer of 1933, David moved to Dartington Hall to take up his duties as science mentor. He entered enthusiastically into the life of the Dartington Hall community. The years at Dartington Hall had a profound influence on Lack, not only as a teacher and a scientist, but also in terms of his socialization as a young adult.

David, who had been an avowed pacifist since the age of 17, found a group of like-minded individuals that included the Elmhursts, Bill Curry, and other members of the Dartington Hall community, who met regularly to discuss their views. After one such meeting that exposed fissures in the group regarding the means of achieving their objectives, David wrote a lengthy "white paper" addressed to Leonard Elmhurst, expressing his personal views on pacifism:

Dear L. K.

At the last meeting of our pacifism group, there was some disagreement among members regarding the question of non-violence. I thought that it might help to achieve common ground if I put my views on paper. This is done partly in the hope that those who hold opposing views will do likewise, since this seems the most satisfactory way of achieving common ground quickly. It was argued by some people at the meeting that we should rather be trying to discuss what methods were expedient to adopt in the present situation than to discuss the moral basis for pacifism. But it seems to me essential to have a sound basic attitude agreed upon before we discuss expedient methods. Otherwise we will inevitably wrangle as to what methods are expedient or justifiable, and much time will be wasted. There seem to me to be two different attitudes towards pacifism, and I am not sure which is held by the group:

1. There are some who wish to stop the present series of international wars by any means whatever (including force of varying degrees if likely to be effective; most pacifists in this group do not advocate bloodshed as a method, though some, eg many communists, do).

2. There are those who consider it desirable that all disputes including international ones be settled by rational means. I myself am definitely in the second group.[10]

The white paper continued with a detailed exposition of six reasons in support of Lack's personal position and proposed means of arriving at a consensus on the issues. A later line presaged David's own response years later, when confronted with the reality of World War II: "[I]t seems to me that one can be consistent in working for peace and yet agreeing to serve the government in the event of immediate war."[11] The paper also presages the form of argument that he used in so much of his later scientific writings, developing clearly articulated optional explanations and then arguing forcefully for his own conclusion.

During his years at Dartington Hall, David made at least two presentations at Sunday Evening Meetings. In November 1935 he presented a talk entitled "The Meaning of Song and Spring Fighting in Birds." After his return from the Galapagos Island expedition (see Chapter 3), he presented a talk in October 1939 entitled "The Disenchanted Islands," apparently reflecting his own disenchantment with his experiences on the expedition. He also made a presentation on East Africa to the entire school after his return from Tanganyika (see below).

David was also involved in sports. He was an active member of the tennis club and the [field] hockey club. The Dartington "News of the Day" reported that the mixed doubles team of Dorothy Elmhurst and David Lack defeated Miss Jewell and Mr. Seyd in July 1935; a year later, Mr. Seyd gained his revenge: "F. Seyd and M. Foos beat D. Lack and M. Richards." David was also an enthusiastic hockey player, one former student recalling that he was "terrifying" on the hockey field, apparently reflecting a fierce competitiveness in David.

One of the experiential components of the progressive education at Dartington was camping trips led by faculty members during the summer term. Lack took a group of students to Lundy Island one summer. Richard Leacock, one of the students, recalled:

We set up tents and Lack took us all over the island observing the bird-life: Puffins, Razorbills, Gannets, Terns, and in the early evening, as it got dark, thousands of Shearwaters who came in and laid their eggs in rabbit burrows. We would run madly among the burrows grabbing the birds and fixing identifying rings on their legs, part of an international attempt to trace their migrations and homing instincts.[12]

During another summer term camping trip, Lack and a second teacher took students to Snowden, at 1085 m the highest mountain in Wales. In June 1938, Lack and Dan Neylan, who taught classics, took a group of students to New Forest, a former royal hunting ground in Hampshire first established in 1079 by William the Conqueror. Peter Thomas, one of the students on the trip, recalled that the group had one bell tent and two or three smaller tents.

One of the most comprehensive descriptions of Lack's teaching at Dartington Hall comes from one of his students, the late writer Eva Ibbotson. In March 2006, she penned an article for *Times Educational Supplement* identifying him as her best teacher:

David—or Lack as we called him, although we could have called him David—
was exceptional. He taught me that the robin was more than a bird; it was a
Marco-Polo-type explorer, and a John Keats of song. David never had any
trouble with discipline, unlike some teachers at Dartington. His lessons were
so clear, and in my writing since I have aimed at a similar clarity. He helped me
to see biology as a fascinating, rich subject; there was none of that awful cutting up of dead animals that was to come when I went to university. He was
also a gifted artist and very musical. He introduced me to *The Beggar's Opera*,
singing all the main parts himself.

I remember going with a few others—or was it just me?—to see him in his
room after he had broken his leg. David was sitting on his bed wearing pajamas,
playing his guitar. You couldn't get away with that now, but at the time this easy
informality seemed perfectly innocent as well as natural. There was a boiled egg
on a plate by the side of the bed, which he ate whole. He then asked us to guess
why he was doing this, eventually explaining that while he normally left the
shell he was consuming it this time because his broken leg needed all the cal-
cium it could get. . . .

It was because of David that I decided to become a biologist. . . . What David
gave me as a novelist was a strong feeling for terrain, which is why I always try
to give my novels a definite sense of place. . . . Outwardly shy with other adults,
particularly women, he found it easier to relate to children. . . . He used to dress
vaguely; corduroy trousers or flannels and sweaters. His eyes and his hair were
pure hazel. . . . But for someone like me, coming straight from Vienna, David
almost was England. He looked so English, and did such English things. Such
a nice man.[13]

Ibbotson's reflections on Lack as a teacher were corroborated by other Dartington
Hall students. Nick Rea (now Lord Nicolas Rea) remembered him as a warm and
imaginative teacher who sensed where students were and taught at their level without being patronizing. Rea, who went on to become a physician, also remembered
that Lack was quick to encourage students' interests, in particular encouraging his
own interest in dragonflies. He recalled that as housefather in one of the senior student houses, Lack would invite several students to his room one evening each week
and would read to them or ask questions to encourage discussion. On one occasion,
recalled Ann Hope Wolff, Lack read James Joyce's description of hell to the students
(presumably that in *A Portrait of the Artist as a Young Man*), no doubt prompting a
lively discussion. She also commented, "I remember seeing him through the window, creeping round the trees observing the robins; this was standard. He was deeply
popular and respected."[14] Another student remembered that on one night in 1936,

Lack, as housefather, rousted all the students out of bed and onto the flat roof of the house to observe a spectacular aurora borealis.

Etain Kabraji Todds, another student, also recalled the evenings spent in Lack's room. Her remembrances are of his singing Elizabethan love songs to his own guitar accompaniment. She also remembered him as an impressive teacher and a person who knitted during his lessons (along with the students, who were knitting as part of the war effort).

Lady Katherine Tait, Bertrand Russell's daughter, remembered Lack as a shy but good teacher. She recalled an instance in which the children in his class placed alarm clocks in several of the cabinets around the classroom, scheduled to go off every 10 minutes. David accepted the practical joke with equanimity.

TRIP TO TANGANYIKA

Julian Huxley had taken note of David's early papers and had become his unofficial "supervisor" while David was still at Cambridge. He recommended that David take a trip to the tropics to balance his experience from his two Arctic expeditions. In response, during his first summer holiday from Dartington Hall School in 1934, David planned a trip to visit Reg Moreau at the Agricultural Research Center at Amani, Tanganyika. Tanganyika, which had been part of Deutsch-Ostafrika before World War I, had become a British colony as part of the Treaty of Versailles. During German rule, the Amani facility had been a biological field station.

David's journey to Amani was itself quite an adventure. He purchased round-trip tickets on Imperial Airways from London's Croydon Airport to Moshi, Tanganyika, for £196 (almost two-thirds of his £300 annual salary at Dartington Hall). Because the planes flew only during daytime, the airfare included accommodation and meals at the overnight stopping points. He would later characterize this transaction as "the best investment I ever made."[15] His travel and accommodation expenses while in Tanganyika were modest compared with the airfare.

Imperial Airways was founded in 1924 with the expressed goal of providing passenger and airmail service to the capitals of the far-flung British Empire. The Empire, though it may have begun to show some cracks at its seams, was still in its heyday, and "imperial" had not yet become a four-letter word. Imperial Airways operated until 1939, when it merged with British Airways Ltd. to form British Overseas Airways Corporation.

The outbound journey was something like a 9-day relay race. On August 1, David flew from Croydon to Le Bourget Airport near Paris aboard the *Hercules*, the first of the Handley Page H.P.42W planes delivered to Imperial Airways in 1931. This was a four-engine, unequal-span biplane with the shorter, lower wing emerging from the fuselage and the engines mounted at the middle of the upper wing. It had a cruising speed of 95 to 105 mph, a range of 500 miles, and it carried a maximum of 38 passengers. Illustrative of the fact that French protectionism is a long-standing phenomenon, the London-to-Paris flight was the only Imperial Airways flight permitted to land in France, and consequently the next leg of the journey was by train, from Paris's Gare St. Lazare to Brindisi on the eastern coast of the boot of Italy. The train left Paris that same day, traveling through Dijon, Turin, Genoa, Pisa (where David glimpsed the leaning tower), and Rome before arriving in Brindisi at 4:00 a.m. on August 3. Passengers were then ferried out to the *Sylvanus*, one of three Short S.17 Kent flying boats that Imperial used on its Brindisi-to-Karachi route. This was also a four-engine biplane with the engines mounted at the center of the upper wing. It had a maximum speed of 137 mph and a range of 450 miles; it carried 15 passengers in addition to a crew of 3.

The *Sylvanus* left Brindisi shortly after 6:00 a.m. on August 3 and made refueling stops in Athens, Mirabella Bay (Crete), and Alexandria before arriving in Cairo just after dark, where David recorded in his journal that he saw the "pyramids silhouetted against the reddened sky."[16] He spent the night in a Cairo hotel and was awakened at 3:15 a.m. in order to reach the airport by 5:00. The next leg of the journey

FIGURE 7: The *Sylvanus* in Athens harbor, 1934. (Photograph by David Lack, courtesy of Lack family.)

took them up the Nile River valley. Refueling stops that day were at Asslut (Asyut) and Assuan (Aswan) before arrival at Wadi Hlifa for lunch (where David also obtained a Sudanese visa). The plane then continued on to Khartoum, where the passengers spent the night in a hotel. Engine trouble delayed the departure from Khartoum, and David reported seeing the juncture of the Blue and White Niles as the plane departed at 8:00 a.m. The plane made a refueling stop, in Kosti, before traveling to Malakal for lunch. Here David reported seeing his first "negroes," Nilotes whom he described as nearly naked, with clayed and trimmed hair, and carrying "formidable-looking spears." The plane then continued to Juba, where it was forced to land after dark "with flares attached to the wings" because of the morning's delay. The hotel in Juba consisted of two-person huts.

The following morning (August 6), the plane arrived in Entebbe, Uganda, at 10:30 a.m., where David got his first view of Lake Victoria. After refueling, it continued on to Kisumu, Kenya, where David changed planes to the *Almathea*, one of the eight Armstrong Whitworth A.W.15 Atalanta aircraft in the Imperial fleet. These were single-wing planes with two engines mounted on each wing, a cruising speed of 130 mph, and a range of 640 miles; they carried nine passengers and three crew members. As the plane flew from Kisumu to Nairobi, where they spent the night, David got an excellent view of the Rift Valley. In his journal he concluded, "[P]hysical geography could be beautifully taught from the air."[17] On the morning of August 7, the *Almathea* reached Moshi, Tanganyika, arriving at 11:00 a.m., and David's outbound flying was over. He remained in Moshi overnight and the next day was driven to Arusha, 50 miles to the west, where he spent the night. He took a long afternoon walk in Arusha and commented, "[W]hen it grew towards evening, the sky cleared and the snows of Kilimanjaro came out of the clouds, looking vast and glorious. The setting sun lit the peak with a pinkish glow, long after it had gone from us, and later it stood out all cold."[18]

David was awakened the next morning at 6:00 a.m. to catch the train from Arusha to Muheza (Mwanza). En route he had a long conversation with a German, whom he described as "an ardent pacifist, agreeing as to the futility etc. of it."[19] But the man was ardently pro-Hitler "and firmly agreed with the expulsion of the Jews." David also observed that there were three lavatories on each station platform—the smallest for Europeans, the next larger for Indians, and "much the largest for natives."[20]

The Moreaus were waiting on the platform at Muheza when the train arrived after dark. Lack later described the man who met him at the railway station in Maneza, Tanganyika, as being "then 37, short, spectacled, nearly bald, not so rubicund as later."[21] The Moreaus drove David and his luggage, along with a 60-pound station mailbag, the remaining 25 miles to the station; the last part of this trip included 10 hairpin curves as they rose to 3000 feet. Along the way, the car flushed two nightjars from the road; this, according to David, was "one of the most impressive incidents of my trip."[22] After 9 days on the road, he had arrived!

David spent most of the next 4 weeks at the Amani station. Moreau introduced him to the variety of tropical habitats near the station, from savannah to montane forest. They took some bird censuses, and David spent parts of many days observing either black-winged red bishops or a pair of Waller's chestnut-winged starlings. These observations resulted in two publications in *Ibis* on the two species (1935 and 1936, respectively). The proposed title on his initial submission to the journal, "Territory and Polygamy in a Bishop," was deemed by the editor to be too risqué; hence its published title, "Territory and Polygamy in a Bishop-bird." In addition to this field work, David also spent time examining the impressive collections that Moreau had amassed, with considerable help from native collectors, and reading some of Moreau's manuscripts, with which he was much impressed. The Moreaus also took him on a 2-day trip to the coast, visiting Tanga and Mwambani.

David's frequent references in his journal to the various native tribes that he encountered reflected something like an anthropological interest, and his attitude toward them seemed to be typical of Englishmen of the day. Of one tribe he encountered he wrote, "These forest dwellers are physically of a low type."[23] In this same village, he "saw the only women with bared breasts on the whole trip." He also commented in his journal,

One wonders what the gain is for all education here, including the government schools. The native is taught to read and write, but then goes home and refuses to labour in the fields and is a burden to his parents, who support him.[24]

He also remarked, "The attitude to religion is exemplified by one of Moreau's servants, a Christian, who wanted a second wife (quite customary in his tribe) and so turned Mahommedan."[25]

The Governor of Tanganyika, Sir Harold MacMichael, made his first visit to the station on August 25. Probably unbeknownst to David, MacMichael had also attended Magdalene College, Cambridge, graduating with a First in Classical Tripos in 1905. He had been appointed Governor in 1933 and served until 1937, when he assumed the difficult post of High Commissioner of Palestine. While still holding that office, he escaped an assassination attempt by the Jewish underground in 1944 in which his wife was seriously wounded. When he visited the station, Lack reported in his journal,

We were all invited to a tea-party on the director's lawn. Rather, but not too excessively formal. The Governor had a fair presence, but, I suppose inevitably, made rather fatuous remarks, as if interested when not specially. I helped Moreau to prepare an exhibition of bird skins for the occasion.[26]

David bid farewell to the Moreaus and Amani on September 7 to embark on a climb to the saddle of Mount Kilimanjaro before heading home. He traveled by overnight train from Muheza to Arusha, then hitched a ride with a lorry from Arusha to Marangu, located near the southeastern base of the mountain. There he met the Rev. Richard Gustavovich Reusch, a Lutheran missionary whom he described as

an extraordinary man. . . . A Cossack by birth, educated at a German university. Of extremely small stature, thin and agile with relatively long legs and a shiny red face. Extremely courteous and deferential in manner, but yet dignified and charming. He must be about 60 years old, and has now climbed to the top of Kilimanjaro 26 times.[27]

Reusch was indeed an extraordinary individual. Born in 1891 in czarist Russia, he served as an officer with the Cossacks during the civil war that followed the Bolshevik Revolution in 1917. Fleeing Russia after the defeat of the White Russians, he studied in Germany in preparation for missionary work and was sent to Tanganyika in 1923. On his first ascent of Kilimanjaro in 1926, he discovered a frozen leopard on the rim of the crater, an event that ultimately helped to inspire Ernest Hemingway (see epigraph to *The Snows of Kilimanjaro*, first published in *Esquire* in 1936). On his second ascent the following year, he discovered the inner cone that now bears his name. He served four terms as a missionary in Africa, ultimately climbing Kilimanjaro 65 times, before retiring to a post as a professor at Gustavus Alolphus College in Minnesota.

Although David had attempted to contact Reusch prior to his arrival, the message had not been received. Because the next day was Sunday, Reusch would not be available to accompany David on his ascent, but he arranged for three African porters. David left Marangu at 1:45 p.m. on September 9 with two of the porters in front and the head porter (and cook) following David and carrying David's suitcase on his head: "We looked grand, but the effect was a little spoilt by my very European and cheap suitcase on the head of the chief porter."[28] The climb to Bismarck Hut at 8500 ft required almost 4 hours, and the party stopped there for the night. Early the next morning, David described the scene from the hut:

The eastern sky was magnificent. Looking that way, one saw a landscape of plains and distant hills, finally the sea with capes, bays and estuaries, with the smoke of ships in the distance, all against a red sky. It was some time before I realized that this whole landscape was composed of clouds.[29]

The second day's trek took the party from Bismarck Hut to Peters' Hut at 11,500 ft. David kept notes on the bird species he observed and on the changing vegetation as they made their ascent. He was impressed by the forests of tree-like heather 30 ft tall and by an arborescent groundsel seen at higher elevations. When the party emerged from the heather forest into grassland, they saw the two peaks of Kilimanjaro clearly, Kibo (19,330 ft) "dome-shaped and covered with snow" and Mawenzi (16,890 ft) "jagged and rocky, with almost no snow."[30] He prepared for the cold night in Peters' Hut by "putting on two vests, two shirts, cardigan, jacket, pants, two pairs of trousers, and 2 pairs of socks and mosquito boots."[31]

The next morning David and the head porter set off for the saddle at 9:10 a.m., reaching it after an hour and 40 minutes. David was showing definite signs of altitude sickness: "Felt pretty rotten, with a bad head-ache and queer stomach."[32] The few remaining bird species gradually disappeared as the party climbed, with the last species, *Pinnarochroa*, dropping out about 200 ft below the 14,500 ft saddle. After a half-hour resting at the saddle, David returned to Peters' Hut, where he again spent the night. The next day he descended to Marangu, reaching Reusch at 3:50 p.m. The cost of the climb to the saddle, including payment of the porters, use of the huts, and food and other supplies, was less than £3. After an overnight stay at the Kibo Hotel, where he "had a shave and bath, and disinfected my feet with potassium permanganate,"[33] David set off on his homeward journey to England to begin his second year of teaching at Dartington Hall.

It took him about 2 hours to hitchhike from Marangu to Moshi, getting lifts primarily from lorries. Late that afternoon, David boarded an airplane for the trip home, essentially the reverse of the outbound journey. On the flight from Moshi to Nairobi, the pilot flew low over a game reserve, and David recorded that he thereby obtained the best views of wildlife that he had on the trip. He described a display made by male ostriches to the low-flying plane: "One went down with its body on the sand and wings spread, looking quite grotesque. . . . It was almost as if the birds were worshipping the strange air-deity."[34] The plane reached Nairobi after sunset. Flares that had been set to illuminate the runway instead caused a grass fire, and "after circling round several times we had a bumpy landing on another part of the ground."[35]

The next day, members of a parliamentary delegation to Uganda joined the flight at Kisumu. At a lunch stop at Entebbe, David walked to the shore of Lake Victoria before the plane resumed its flight to Juba for the night. Most of that flight was at 10,000 ft, but the pilot dropped in altitude at Murchison Falls, circling the falls four times. The next 2 nights were spent in Khartoum and Luxor. After dining in Luxor, they visited the temple of Karnak:

We first sailed down the Nile, with the felucca sail, with a glorious golden moon and silhouetted palms, and a balmy air; it would have been lovely if it had not been laid on so thick and if the boatman had not been paid (obviously) to sing a boatman's song. Anyway it would have been quicker to have driven by car rather than sailed, and we did return by that method.[36]

Eminently practical, but hardly the soul of romance.

Two more days of flying took David from Luxor to Brindisi, where he caught a train to Paris, arriving there some 36 hours later, at 7 a.m. on September 20. The flight from Le Bourget lasted two and a half hours, reaching Croydon at noon. "Over the Channel the plane met a rain storm, there was a huge bump, things flying from the racks, and all of us leaving our seats. The steward said we had dropped a thousand feet, the biggest bump the pilot had ever known."[37] It was a fitting conclusion to a remarkable adventure of discovery.

FIRST TRIP TO THE UNITED STATES

During the summer holiday of 1935, David made his first trip to the United States, serving as a chaperone for a student returning from Dartington Hall to California. He boarded the appropriately named *SS Manhattan* at Southampton on August 1 and sailed into New York harbor in the early morning of August 8, after an uneventful crossing during which he read T. E. Lawrence's *Seven Pillars of Wisdom* and played numerous games of ping-pong. He was picked up at the dock by Beatrice Straight, who drove him to her apartment near Central Park, where he saw his first two land bird species in North America, both European interlopers–the house sparrow and the starling. Although a brief romantic interest had apparently developed between Beatrice and David during David's first year at Dartington Hall (she was only 4 years younger than the handsome 23-year-old biology master), it had ended when she came to New York to pursue her acting aspirations. She had made her Broadway debut in *Bitter Oleander* earlier that year.

Later that day, "P"[38] and her sister, a student at Vassar College, picked David up at the apartment in a new Ford V-8 to embark on their transcontinental journey to the girls' home in Berkeley, California. Their route westward followed the Lincoln Highway, the first transcontinental highway across the United States, which ran from New York to San Francisco. They traversed Pennsylvania, Ohio, Indiana, Illinois, Iowa, Nebraska, Wyoming, Utah, and Nevada, finally crossing the infamous Donner Pass in the Sierra Nevadas of California to reach the Bay area. They arrived in Berkeley on August 14, and David stayed there with the family for several days.

David began his journal entry for August 13 with the words, "Indeed unlucky."[39] They had had two automobile accidents that day, both in Utah. In the first, another car struck theirs from behind, then veered into a trailer, knocking it from the highway and "strewing the contents all over the road." The Ford was only slightly damaged, and they proceeded westward, only to back into a passing car when they were leaving a parking lot after stopping for a sandwich. Wrote David, "Large dents in her side, and our carrier with my suit-case smashed up. Nothing serious."[40] Again they were able to continue their trip after resolving liability issues with the other driver.

David's first stop was the Museum of Vertebrate Zoology of the University of California, Berkeley. There he met the pioneering ornithologist/ecologist Joseph Grinnell, as well as ornithologist Robert McCabe and wildlife biologist Lowell Sumner. Lack stayed more than a week in the Berkeley area, spending most of his time birding, sun-bathing, and playing tennis. On August 22, he spoke to the Cooper Ornithological Club on the topic, "Recent Observations of Territory in Bird Life," with Grinnell and Alden Miller among the listeners.

David left Berkeley on August 24 to visit other parts of California with traveling companions Laidlaw Williams, a young ornithologist, and Mrs. Lowell Sumner. They spent 2 days in Carmel, then drove across the Coast Range and through the San Joaquin Valley into Yosemite National Park, where they camped for 3 nights. The last night was spent at the Sierra Club camp at Soda Springs, where they "added whiskey to the natural soda water from the spring."[41] Lack returned to Berkeley on August 30 to discover that his room was in total disarray due to efforts to remove a large swarm of bees from under the roof, an event that was unusual enough to merit coverage in the Oakland press. A travel agency in Berkeley helped him prepare his itinerary for his return trip to New York by bus, with numerous stops and side trips. He left Oakland at 6:30 p.m. on August 31 for an overnight trip to Los Angeles and changed buses there for Indio. There he was met on the morning of September 1st by Benjamin Clary, whose wife, Marjorie, was an amateur botanist; Clary drove him to their home, Coral Reef Ranch, near Coachella.

The following morning, Benjamin drove David to the Salton Sea, which lies in the floor of a rift valley on the San Andreas Fault. They spent 2 days at the lake, which is the largest in California and whose surface lies more than 200 ft below sea level. The lake is surrounded by extremely arid desert, and the temperature on September 3 exceeded 120°F. They visited the mud pots, which Lack described as "boiling mud springs, the water and mud coming up in great globular gloops."[42] At one, a geyser suddenly shot 100 ft in the air near where he was standing. After driving around the Salton Sea, Clary delivered David to the evening bus leaving from Indio.

He stopped in Cedar City, Utah, where he had arranged for side trips to Bryce and Zion National Parks and the Grand Canyon. He described Bryce as a great semi-circle of bare rock, surrounded by green pines, which stop abruptly on the edge. Within the semi-circle a mass of great red and white pinnacles on every side, in regular rows, like a celestial city, and beyond a huge plain broken up by scarps and cliffs again in red and white, to the setting sun; a quite indescribable effect.[43]

He was quite unimpressed by the banality of the remarks made by Americans viewing the wonder at the same time. Two that he recorded were, "Pretty, isn't it?" and "Heck of a place to lose a cow in." At the Grand Canyon, he took the Kaibab Trail to the floor of the canyon, leaving at 5:45 a.m. The trail followed a mule path along Angel Creek, which had to be forded six times during the descent. He reached the floor before noon, lunched there, and began the climb back at 12:15 p.m. He was disappointed that he could not see either rim from the floor of the canyon, and he described the Colorado River as "very muddy." He arrived at the rim at 7:15 p.m., quite exhausted from the 30-mile hike, which had taken him on a vertical descent from 8100 to 2500 ft and back again.

On September 9, Lack boarded a bus at Cedar City headed for his next destination, Denver, Colorado, which he described as "an unpleasant coal town."[44] From Denver he took a 2-day side trip to Rocky Mountain National Park. He returned to Denver late on the second day and, while waiting more than 6 hours for the bus eastward, encountered yet another American institution in the persons of two grifters. The first engaged him in conversation and persisted despite David's efforts to lose him. When David and his new-found acquaintance walked out of the bus station,

> an extraordinary person accosted us in an American-talkie idea of an Englishman's voice, saying could we tell him the way to the Anglo-American Club. My 'friend' engaged him in conversation, and the 'Englishman' embarked on a long story of how he had been here only a few days, having come to collect the fortune left by his brother, who had died of delirium tremens. The man's accent was only a little overdone, his vocabulary perfect except that he once used the word 'phoney,' and then embarked on a long story about a four-pound note'. . . . Obvious confidence men.[45]

David managed to disengage from them and boarded his eastbound bus. The trip from Denver took him through Kansas City, Chicago, and Cleveland before he arrived in New York, after 72 hours of continuous travel, at 11:30 p.m. on September 16.

On the morning of the 17th, he went to the American Museum of Natural History, where he met Ernst Mayr, who became a lifelong friend and frequent correspondent.

Mayr showed him the "magnificently housed" Tring collection of bird specimens amassed by the famed British collector, Lord Walter Rothschild. At the museum he also met Frank Chapman and G. K. Noble, and in the evening he attended a dinner with the New York Linnean Society before catching a taxicab to board the *SS Bremen* at 11:30 p.m.

The *Bremen*, which was operated by the German firm Norddeutsche Lloyd, had been launched in 1929 as one of a new class of luxury ocean liners, and on her maiden voyage to and from New York had broken the records for trans-Atlantic crossing times in both directions. She had a catapult between her stacks that could launch an airplane, and on that voyage she had launched a Heinkel HE 12 aircraft to deliver mail to New York before landing. In July 1935, a few weeks before Lack boarded her for his return trip to Southampton, the *Bremen* had been boarded in the New York harbor by a group of anti-Nazi protesters who tore the Nazi flag from the flagstaff and tossed it into the Hudson River. This prompted Adolph Hitler to proclaim on September 15, 1935 (2 days before Lack's boarding), that the Nazi flag was the exclusive national flag of Germany.

Lack was booked third class for the voyage and noted that he was one of only four Englishmen on the third class passenger list. Almost all of the rest were German, and he "was much struck by the dancing (nazi style), very military and no jazz, mostly one-steps and the Viennese waltz, very severe."[46] The ship arrived at Southampton at noon on September 23, and from there David took a train to Totnes and Dartington Hall for another year of teaching. His first trip to the United States had cost him exactly £72, and he had observed 254 bird species, 30 of which he had previously seen in Britain.

DARTINGTON HALL ENCORE

After arriving back in England in late September 1939 from his Galapagos expedition (see Chapter 3), David returned to Dartington Hall School for what was to be his last year of public school teaching. He and Bill Hunter, the film teacher at the school, used the motion picture footage taken by Richard Leacock in the Galapagos to make a short silent film of the creatures and habitats of the islands. He also continued to work on his data from the Galapagos and from the museum specimens that he had measured, and he completed the manuscript of a major paper on the Galapagos finches that he submitted to the California Academy of Sciences in May 1940.

During his last year of teaching at Dartington Hall, the long shadow of the unfolding war in Europe began to fall on the school. The London County Council evacuated students from three primary schools to Dartington Hall to protect them from the London blitz. At the same time, because the east coast of Devon was thought to be one of the more probable invasion routes for Hitler's armies, parents were pulling their children out of Dartington Hall to ensure their safety, and both teachers and senior schoolboys were leaving to join the British armed forces. As one student described it, "Every day trunks were piled up in the entrance hall and there were sad goodbyes to be said."[47] David, who was to leave the school for good at the end of the 1940 summer term, expressed the mood in Miltonic verse:

> Weepe, daughters of Dartington, Weepe ye and waile!
> Sing Tragick Muse of dear departed Male.
> O Dartington, your glorious Sonnes are fled . . .
> Some home, some Interned, some Banished
> To distant Lands (tho' theirs was not the crime).
> Our Patrick he is gone, gone ere his prime.
> Bryan now bathes in shark-infested cove
> Or wanders listless in some orange grove.
> O Muse, enquire where are the Songs of Spring?
> Where Kevin's shout and Dick's quainte murmurings?
> Walter no longer hums his basso softe
> No Danny whistles from his Roof aloft.
> Goldstein no more complaynes upon his Flute
> Even our Bell is mute.[48]

Many, like Lack, never returned to Dartington. "Kevin's shout," which referred to a senior school student, Kevin Black, was silenced forever on D-Day. Bill Hunter, the geography teacher who founded the Dartington Film Unit and collaborated with Lack in producing Richard Leacock's Galapagos film, died in 1941 while serving in the R.A.F.

WORLD WAR II

David was still a committed pacifist when he left Dartington Hall School in the late summer of 1940. He had signed the pledge of the Peace Pledge Union ("I renounce war, and I will never support or sanction another"), and he joined a pacifist unit working in the East End of London. He found the "earnest attitudes" of his fellow pacifists off-putting, however, and one night there experiencing firsthand the horrors of the London blitz converted him from pacifism. He soon responded to an invitation from the Central Register for Scientific Workers to join a unit working on the newly developed wartime uses of radar. After training in Richmond, his first assignment was to a heavy antiaircraft battery in the Orkney Islands, where, as he said, "I had a wonderful holiday from March to August 1941."[49] He had 1 day off each week, which he spent birding on the other islands.

A baker in Kirkwall named George Arthur was an avid bird watcher and local representative of the Royal Society for the Protection of Birds. His home served as a meeting place for ornithologists serving in the islands, and they often went bird watching in his baker's van, after which he would provide a "tremendous" meal. David clearly found him most charming and helpful: "[H]e repaired everyone's wireless, played the cello in the city orchestra, was an expert bird-watcher and fly-fisherman, and exuberantly hospitable to all ranks of both sexes."[50]

When he returned to Richmond in August, David was assigned to the Army Operational Research Group (AORG) headed by Lt. Col. B. F. J. Schonland. Schonland was a South African professor of atmospheric physics who had been elected a Fellow of the Royal Society in 1938 and who went on to win the South Africa Gold Medal of the Southern Africa Association for the Advancement of Science in 1943. He was knighted in 1960. Lack first served as Schonland's personal assistant for 6 months, spending much of his time abstracting complex data sets to provide Schonland with compact briefings. Although he found that he was proficient at the task, it proved so demanding that he was exhausted at the end of the 6-month period and was relieved to be assigned to the field.

He spent the remainder of the war traveling among radar installations, performing various duties such as checking the effect of weather on radar propagation. He watched the invasion of Normandy on D-Day by radar from the installation on Beachy Head (East Sussex). On a trip to installations on Orkney and Shetland in February 1945, he contracted dysentery, and the return flight to the mainland ended with a nose-down crash landing. The combination of the crash and the dysentery nearly killed him.

One of the principal tasks of field-based AORG scientists was to support the land-based radar stations that were responsible for tracking ships and aircraft along the coasts of Britain. While serving in the field, Lack had the opportunity to interact with numerous other young biologists, and he reported that the tea-time conversations were very stimulating. One such young scientist whom he had known as a contemporary at Cambridge was the entomologist, George Varley. Varley, who was later to be named Hope Professor of Entomology at Oxford, introduced Lack to the importance of density-dependent regulation of animal populations, an idea propounded by the Australian entomologist A. J. Nicholson that would have a profound influence on David's later work (see Chapter 5).

FIGURE 8: Crash-landing of airplane on which David was a passenger during World War II. (Photograph courtesy of Lack family.)

One of the problems that plagued wartime radar operators was the occurrence of spurious echoes, which often were referred to by operators, most facetiously but sometimes seriously, as "angels." In February 1945, Lack prepared AORG Report No. 297, entitled *Radar Records from Birds*, a classified report over Col. Schonland's signature. Referring to the spurious echoes, he wrote:

> The two most puzzling features of such echoes are first that they often move across or against the surface wind, and secondly that when an operator makes a visual observation of the radar range and bearing, he usually reports that there is nothing there. . . . Undoubtedly such unidentified echoes are due to a multiplicity of causes, and this must always be kept in mind. However, it is the opinion of A. O. R. G. first that the great majority of such echoes are due to birds, and secondly that a study of the signal-strength will normally distinguish bird echoes from those given by ships, aircraft or other objects of operational importance.[51]

In this four-page report, Lack described the three main objections to the conclusion that most of the spurious echoes were caused by flying birds: (1) the radar signal from birds is too weak; (2) birds don't fly at night ("The latter objection is usually made by scientists who are not ornithologists"); and (3) when an observer searches for objects at the range and bearing of the spurious echo, there is nothing there. He described several direct confirmations of detection of flying birds by radar, including the first such observation at Dover in September 1941, when an observer using a telescope identified flying gannets "at radar lengths as great as 15,000 yards," corresponding to the simultaneous observation of "spurious echoes" at that range and bearing. The observer making the visual identification of the gannets was later identified as George Varley. Lack also described other visual confirmations of radar detection of flying birds, including those of gulls observed at Gibraltar, a white stork observed on Malta, and European starling flocks flushed from their nocturnal roosts by incoming V-2 rockets.

After the war, Lack and Varley co-authored a paper in *Nature* (1945) describing the discovery that flying birds are detected by radar. The other main scientific contribution arising from the wartime observations was documentation of the frequency and size of nocturnal flights over water made by birds. In the same year, an American ornithologist published a similar account in *Science*. In August 1999 the BBC produced a fictionalized account of Lack's and Varley's radar discoveries entitled *Wings of Angels* (see Chapter 4).

While working in the field for AORG, Lack spent many evenings and Sundays writing. He completed the manuscript for *The Life of the Robin*, which was only half-written at the outset of the war, in 1942. The following year, he began work on a book for sixth form students based on his observations on the Galapagos Islands, a work that evolved into *Darwin's Finches* (see Chapter 3). He was also able to attend some professional meetings, most importantly presenting a paper to a symposium on the biology of closely related species at the annual meeting of the British Ecological Society in March 1944 in London (see Chapter 3).

Lack was released early from the AORG, in August 1945, because he had been appointed Director of the Edward Grey Institute of Field Ornithology at Oxford (see Chapter 5).

[T]here is no evidence whatever, in any of the island forms of Geospizinae, that their differences have adaptive significance.

—*The Galapagos Finches (Geospizinae), A Study in Variation*, p 117

My views have now completely changed, through appreciating the force of Gause's contention that two species with similar ecology cannot live in the same region.

—*Darwin's Finches*, p 62

3

Darwin's Finches

THE ASSERTION THAT David Lack is the father of evolutionary ecology rests primarily on two works published in 1947. The first of these, *Darwin's Finches*, was a book based on Lack's studies in the Galapagos Islands before the war. The second was a lengthy paper published in three installments in *Ibis* (the third installment actually appearing in 1948), entitled "The Significance of Clutch-Size." The latter is the most frequently cited paper ever published in *Ibis*. Before discussing these seminal works, I will describe the Galapagos expedition that proved to be the fulcrum on which the major change in David's professional life hinged.

GALAPAGOS ISLANDS EXPEDITION

David's interest in territoriality in birds, manifested in his review (co-authored with his father) of the subject in *British Birds*[1] and further stimulated by his findings on the robin, led him to seek a sabbatical from Dartington Hall School to pursue the subject further. His intention was to study territorial behavior in a closely related group of birds. He first considered working on weaverbirds in Africa, but after reading P. R. Lowe's study on the Galapagos finches,[2] and at the urging of Julian Huxley, he decided to work instead on the geospizine finches of the Galapagos Islands. Huxley helped acquire major grants for the expedition from the Royal Society and the Zoological Society of London (of which he was Secretary). Lack selected a student studying film making at Dartington Hall, Richard Leacock, to accompany him on the expedition to photograph the finches. He also chose L. S. V. "Pat" Venables to assist in observations. He had met Venables while working on a heathland census organized by the British Trust for Ornithology. The expedition was initially planned for 1937–1938, but a sudden illness contracted by Venables resulted in a year's delay. Fortunately for David, Dartington Hall had yet to fill his position for the year, and he returned there to teach. It was one of the few fortuitous events to mark the expedition, which proved to be something of a cock-up from beginning to end, including David's differing interpretations of the results. Yet the major work that ultimately resulted, *Darwin's Finches*, is undoubtedly the work for which he is most widely known.

By the time the expedition finally got under way in the fall of 1938, three others had been added to the team. The three, recruited by Venables, were W. Hugh Thompson, an ornithologist, and Dr. and Mrs. T. W. J. Taylor, botanists. Although each was to contribute significantly to the expedition, Lack concluded many years later that "we proved an unwieldy party, and the age range from 17 to 40 was too wide, so things did not go happily, except when we were in groups of two or three."[3]

In actuality, Thomas Weston Johns Taylor was 43, and he was a chemist by training. His undergraduate career at Brasnose College, Oxford was interrupted by service in the Essex Regiment both in France and at Gallipoli during World War I. After completing a First Class degree in chemistry in 1920, he was elected a Fellow of Brasnose College, and he became a university lecturer in organic chemistry in 1927. He married Rosemund Georgina Lloyd in 1932, and it was her interest in botany that spurred Taylor to develop his own passion for the subject. By the time he participated in the Galapagos expedition, he had acquired an encyclopedic knowledge of the flora of many parts of the world. When World War II broke out, Taylor joined the chemical branch of the Royal Engineers and served in the Middle East, in Southeast Asia, and in Washington, D.C., as Director of the British Central Scientific Office. After the war he was named the first Principal of the newly established University College of the West Indies (Mona, Jamaica). He held that post until 1952, when he left to become Principal of the University College of the South West (now the University of Exeter). He received the rank of CBE (Commander of the Order of the British Empire) in 1946 and was knighted in 1952. He died the following year.

At the other end of the age spectrum was Richard (Ricky) Leacock, who was indeed 17 at the time of the expedition. Ricky was a student at Dartington Hall School who was deeply interested in photography and film making. Although he was born in London, he lived on his father's banana plantation in the Canary Islands until he was sent back to England for boarding school. He had read Darwin's *Voyage of the Beagle* as a schoolboy and was intensely interested in the Galapagos. He eagerly accepted Lack's invitation to accompany the expedition as photographer. Afterward, he matriculated at Harvard to study physics and then pursued a career as a pioneering documentary film maker. He also helped to found the Film Unit at the Massachusetts

Institute of Technology in Cambridge, Massachusetts, in 1969, and he retired from his teaching post there in 1989. He was living in Paris at the time of his death in 2011.

David embarked from Liverpool aboard the *SS Reina del Pacifico* on November 3, 1938, arriving in Bermuda 10 days later after a rough crossing. During the 3-hour stopover, he visited the zoo to see the Galapagos penguins and giant tortoises that had been obtained by the Astor expedition to the islands in 1930. He also concluded that the house ("English") sparrow was the most common species on the island. The *Reina* reached Nassau, the Bahamas, on November 15 and Havana, Cuba, a day later. David spent most of a day touring Havana and seemed quite impressed, particularly with the Church of Our Lady of Mercy (Lourdes) and the new, $20 million Capitol. He also observed, in an unusual reference to political matters:

> The current dictator assumed power, after considerable bloodshed, in 1933. Am told he is very left in tendency, has elaborate laws to protect the dismissal of workmen, and has introduced a great deal of education among the poorer people, even in remote villages, all over the island.[4]

> The dictator to whom he was referring was none other than Fulgencio Batista, who would again be in power when the Cuban Revolution led by Fidel Castro overthrew his regime in 1959. Ironically, Batista was supported by Cuban Communists in the 1930s, presumably because of his protection of workers and his support of universal education.

The *Reina* sailed from Havana on the morning of November 17 and arrived in Kingston, Jamaica, at noon the next day. David spent only a short time in Kingston, describing it as "dirty, dull" and decrying the persistence of the street peddlers. The next day the *Reina* arrived at Cristobal, at the Caribbean end of the Panama Canal, which it passed through on November 20. Two days later, David disembarked at Salinas, Ecuador, and boarded a bus for an all-day trip across the deserts of western Ecuador to Guayaquil.

On his first day in Guayaquil, Lack learned that the ship on which he had arranged passage to the Galapagos, the *Jose Cristoal*, was not seaworthy. He therefore began a frantic search for alternative transport to the islands. After several failed attempts to locate transport, he found the *Boyaca*, which sailed from Guayaquil on the afternoon of December 8, with Venables, Leacock, Thompson, and Lack aboard (the Taylors did not arrive until late January). The *Boyaca* arrived in Wreck Bay, Chatham (Isla San Cristobal), on December 14, and the team disembarked with all of their supplies and equipment. Chatham had also been Darwin's first stop in the Galapagos in 1835.

The base camp on Chatham was at El Progresso, a plantation owned by Mr. and Mrs. Cobos. Señor Cobos, a descendent of Manuel Cobos, who had founded El Progresso in 1869, proved to be a much more hospitable host than his forebearer. Manuel had initially established El Progresso to harvest orchilla, a lichen used in making dyes, but he soon developed it as a sugarcane and coffee plantation with the assistance of prisoners in a penal colony that was moved there from Indefatigable (Isla Santa Cruz). Ruth Rose provided a colorful description of the early history of El Progresso in William Beebe's *Galapagos World's End* (G. P. Putnam's Sons, 1924).

Manuel ruled Chatham as his personal fiefdom and treated the prisoners essentially as slaves. In 1904, a sloop arrived in a Colombia port carrying 77 men and 8 women. Because of their lack of papers and other suspicious circumstances, they were detained by Colombian authorities, whose questioning gradually elicited the true circumstances behind their mysterious arrival in Colombia. The party was composed of prisoners from El Progresso who had killed Manuel Cobos and escaped Chatham aboard his sloop, which they had hopefully renamed *The Liberty*. They described the terror under which they had lived at El Progresso, having watched as several fellow prisoners were flogged to death or summarily shot by Cobos, while others were taken to nearby uninhabited islands and left to die there with little water, food, or supplies. Cobos had banished one man, an Ecuadorian soldier named Camillo Casanova, to Indefatigable 3 years earlier, leaving him with only a small container of water, a kitchen knife with no point, and a machete. Ecuadoran authorities sent a ship to Indefatigable and discovered that Casanova had miraculously survived his years of enforced exile on the island, a true-life Robinson Crusoe. Others who had been banished were not so fortunate.

David had no such ill treatment to complain of, but he did record that his first night on Chatham was "extremely cold" despite his blanket, Macintosh, jersey, pullover, and wool shirt. No doubt due to the cold, the mosquitoes were only "mildly unpleasant," a situation that would change significantly later in the expedition, resulting in numerous journal references to poor sleep caused by swarms of mosquitoes.

The next day, David located transportation for Venables and Leacock to South Seymour (Baltra) aboard the *Trondjem*, a modified Norwegian lifeboat owned and captained by an Englishman, Peter Browning, who was accompanied by his wife and two crewmen. Pat and Ricky left on December 17 and spent the next 3 weeks on South Seymour, returning to Chatham on January 9 (also aboard the *Trondjem*). Ricky recalled that he had actually feared for his life on one of the passages when Peter Browning discovered Pat and Mrs. Browning in a compromising position one night. Ricky feared that Pat and he might be mysteriously "lost at sea," a scenario that fortunately did not materialize.

David set about organizing his base camp at El Progresso and soon began a regular routine that he maintained throughout the expedition. He began his days with 3 to 4 hours of field observations on the finches and then spent his afternoons building traps and aviaries for them. Thompson also devoted most of his time to studying the breeding biology and behavior of the finches. On Sundays, David often spent the day exploring some part of the island, usually searching for species of birds that did not occur around his base camp.

Two days after Venables and Leacock returned from South Seymour, Ricky and Hugh left for Indefatigable aboard the *Trondjem*. Upon arrival there, they set up camp on a plantation owned by the four Angermeyer brothers–Gus, Fritz, Karl, and Hans. The four, along with Hans's Dutch wife, Lizzie, had immigrated to Indefatigable from Germany 2 years earlier, at the urging of the Angermeyers' parents, who foresaw the coming of war in Europe and feared for their sons' safety. Their fears proved to be prescient: three of the boys survived World War II (Hans died of tuberculosis in Ecuador during the war), but their parents did not. Ricky formed close friendships with the Angermeyers, particularly with Gus, Fritz, and Karl, and spent a considerable amount of time helping them construct their house. He later said that Lack was probably disappointed that he did not spend more time filming the finches. David's only comment on the matter was a journal entry written January 31, when he arrived on Indefatigable: "Found R[icky] had nowhere near finished his filming, so he stays on, which will make the food problem and sleeping more complicated."[5]

On January 26, 1939, the Taylors arrived on Chatham aboard the *Boyaca*, and the next day David made preparations to leave with them for Indefatigable. On the evening of January 30, the three of them sailed out of Wreck Bay aboard the *Boyaca* to join Ricky and Hugh on Indefatigable, leaving Venables behind at El Progresso. The ship stopped at Hood (Espanola) early the next morning, and the Taylors and David went ashore for most of the day. There David found a Galapagos hawk perched in a tree that showed no response when he tickled its back and head with a stick. He also observed one of the ground finches (*Geospiza conirostris)* that he had not yet seen. The *Boyaca* sailed from Hood that night and reached Academy Bay, Indefatigable, the next day.

David and the Taylors set up camp at the home of Karl Kubler and his wife, Marga. Kubler, who was born in Alsace-Lorraine, had lived in Spain for several years before immigrating to the Galapagos with his wife and his daughter, Carmen, in 1934. He owned a small plantation in the rain forest above Santa Cruz, where he grew bananas, maize, sugar cane, tobacco, and papayas.

David resumed his regular routine of observing the finches for several hours each morning but also undertook several excursions to see new birds (e.g., flamingoes in Tortuga Bay) and the rain forest and grassland/bracken habitats at higher elevations on Indefatigable. One such excursion, on March 2, took him to the top of "Zucker hutte" from which he could see several of the other islands, from tiny Gardner to the southeast to James (Santiago) to the northwest. In his journal, he drew a map of the positions of the islands that were visible from Zucker hutte, including Daphne Major lying almost directly north. Many years later, Daphne Major would become one of the primary research sites for Peter and Rosemary Grant and colleagues for their long-term studies of Darwin's finches. David also located a canister containing notes from the Milwaukee Museum expedition, which had reached the site on January 13, 1933.

Hugh Thompson contracted a serious case of dysentery in mid-February. He remained seriously ill, almost dying according to a later account by Lack, until he and Ricky left aboard the *Seven Seas* to return to Chatham on March 10. David and the Taylors remained on Indefatigable, and David continued to make regular observations on the finches, including performing recognition experiments with stuffed birds. He also captured a number of finches to take back to England for hybridization experiments.

Throughout the team's stay on Indefatigable, Kubler, and occasionally the Angermeyer brothers, augmented their rations with bananas, papayas, and other produce from their plantations. Kubler was also an avid hunter, and he frequently provided fresh meat, killing wild pigs, tortoises, and sea turtles and on one occasion spearing a "pike-like fish." On February 11, Ricky filmed Kubler shooting and cutting the throat of a pig.

Pat, Hugh, and Ricky arrived on April 1 from Chatham aboard the *Deborah*, the ship that was to take the team to Panama—a trip that David recorded cost $1000 US. The *Deborah* sailed from Indefatigable in the afternoon of April 2 with the entire team aboard, as well as the captive finches: 12 *Geospiza scandens*, 6 *G. magnirostris*, 1 *G. fortis*, and 11 *G. fuliginosa*. The *Deborah* anchored in Darwin's Bay, Tower (Genovesa), the next morning, and the team went ashore for a few hours; while there, Ricky filmed the displays of the magnificent frigate-bird. That afternoon, the *Deborah* resumed its voyage to Cristobal, Panama, and it was farewell to the Galapagos.

The first order of business for David in Cristobal was to decide on the disposition of the captive finches. They were in poor condition, and he feared that they would not survive the trip to England. He therefore wired Julian Huxley on April 9, describing the condition of the birds and suggesting that they accompany him to San Francisco instead. Huxley approved this suggestion in an April 11 wire, and David began seeking passage for himself *and* the finches to San Francisco. This task proved to be somewhat problematic, because most ships willing to take him to San Francisco were unwilling to also take the finches.

The team members remained for various lengths of time in Cristobal until they could arrange passage to their ultimate destinations–Ricky to the United Sates to begin undergraduate studies at Harvard; Pat, Hugh, and the Taylors back to England; and David to San Francisco. Ricky left on April 10 for New York aboard a Grace Line ship, and the Taylors left the next day. Hugh and Pat sailed aboard the French ship, *Colombia*, on April 16. The following day, David finally booked passage for himself and the finches aboard the *Peter Maersk*, which sailed from Cristobal on April 18.

The *Peter Maersk* arrived in San Pedro, California, on April 27, where David was met by James Moffitt, the curator of birds and mammals at the California Academy of Sciences. An overnight train ride brought Moffitt, Lack, and the finches to San Francisco, where David turned the birds over to Dr. Paul Kinsey. "Was not sorry to see the last of the cages," he wrote.[6] David now began his 4-month sojourn at the California Academy.

Many of David's journal entries during his Galapagos adventure were less than positive. He frequently complained about the clouds of mosquitoes and fleas, the "jiggers," the weather, and the impenetrable thorn-scrub of the lowlands. On Chatham, he also complained about disturbed sleep caused by raucous nocturnal parties held at the Cobos' plantation. On New Year's Day, 1939, he reported, "V[ery] disturbed night as most of the officers called on the Cobos about 10:40 p.m. & stayed till after 3 a.m. drinking in the New Year with song, gramophone & dance."[7] The end of Christmas festivities on January 6 elicited a similar entry: "For us, each festivity means a disturbed night."[8]

The isolation that the team had experienced in the Galapagos meant that David had only limited and belated knowledge of the events that within a few months were to plunge the entire world into mortal conflict and so dramatically alter not only the course of history but also his own life over the next few years. A March 26th journal entry records that he heard from the Commandante, who had only that day arrived on Indefatigable from Chatham, that war had been declared 9 days earlier between Poland, Hungary, and Czechoslovakia. An April 8th journal entry in Cristobal demonstrates that David was unaware until then of the true gravity of the crisis in Europe: "At last got the European news, which is pretty staggering, Hitler having got Czechoslovakia and Memel [East Prussia], and threatening Poland, Mussolini having got Albania!"[9]

David's journal contains few observations about the people living in the Galapagos. At one point he commented, "For those who like the simple life, here it is."[10] He also noted, on learning of the death of Captain Boysalt's 5-month-old daughter, "The knowledge of hygiene here is pretty lacking."[11]

The lives of the Kublers and Angermeyers remained intertwined after David's departure from Indefatigable. Within 2 years, Marga left Karl and took her daughter

with her to live with the Angermeyer brothers. She eventually married Karl Angermeyer, and her daughter married Fritz. Fritz and Carmen's son, Fiddi, later became
the founder of Angermeyer Cruises, owner of the Angermeyer Point Restaurant in
Puerto Ayora (Isla Santa Cruz), and a pioneer in promoting ecotourism in the
islands. Hans' wife left the islands soon after David's departure to return to the Netherlands, and Hans subsequently married an American widow, Emmasha Aguirre.
Hans and Emmasha were forcefully separated after the bombing of Pearl Harbor,
because all American civilians living in Ecuador were repatriated to the United
States. Hans remained in Ecuador, where he died of tuberculosis shortly before
Emmasha gave birth to their second daughter, Johanna. In 1989, Johanna Angermeyer, by then an artist and writer of children's books living in England, published a
compelling book about her Galapagos heritage.[12]

CALIFORNIA ACADEMY OF SCIENCES

Founded in 1853 as the first scientific academy west of the Atlantic seaboard, the
California Academy of Sciences had lost most of its specimens in the great earth-
quake of 1906 and had relocated to Golden Gate Park in 1916. By 1939, it had ac-
crued a long history of work in the Galapagos and housed an extensive collection of
geospizine finches. Indeed, Harry Swarth, the previous Curator of the Department
of Ornithology and Mammalogy at the Academy who died in 1935, had published
his work on the taxonomy of the finches based on this and other large collections,
and this work had contributed to David's decision to pursue his own research on the
Galapagos finches.

On arriving at the Academy in April 1939, David began analyzing his field notes
from the Galapagos and measuring specimens of the finches. In addition to evalu-
ating the many specimens at the Academy, he also visited the Museum of Vertebrate
Zoology at the University of California, Berkeley, and the Stanford University Museum of Natural History. For each specimen, he measured the length of the wing and
the length and depth of the beak and recorded the sex, plumage coloration, and
collection location. Over the next year, David would measure almost 6400 finch
specimens, including 180 Cocos Island finches, and, at the British Museum, the first
specimens of finches collected by Charles Darwin in 1835. He also measured several
hundred specimens of the house sparrow at the American museums he visited, and
he wrote a short paper on geographic variations in that species occurring in North
America during the first 90 years after their introduction to the continent.[13]

While he was at the Academy, Lack also made the acquaintance of a young ornithologist in the Department of Zoology in the College of Agriculture at the University of

California, Davis, John T. Emlen, Jr. Emlen had obtained a Ph.D. at Cornell University under the supervision of the pioneering American ornithologist, Arthur A. Allen. Emlen and Lack collaborated on a study of the breeding behavior of the tricolored blackbird (*Condor*, 1939).

After 4 months at the Academy, Lack traveled across the United States to New York, where he stayed with Ernst Mayr while engaged in measuring finch specimens at the American Museum of Natural History. He discussed his work extensively with Mayr, and Mayr's influence on his thinking is quite apparent in Lack's later works. One irony of this sojourn was that the two men—Lack the Englishman and Mayr, still a German citizen—were together on September 3rd when they heard that their respective countries were at war. Later that month, Lack returned to England to resume his teaching post at Dartington Hall School.

Despite all of the other problems associated with Lack's Galapagos expedition, however, the most glaringly obvious cock-up was probably his interpretation of its findings.

Darwin's Finches

Lack's field work in the Galapagos and his subsequent examination of specimens of the geospizine finches in several museums, particularly in the extensive collections at the California Academy of Sciences, Stanford University, the Museum of Comparative Zoology (Harvard), the American Museum of Natural History, the U.S. National Museum, and the British Museum of Natural History, culminated in two major publications. Embarrassingly, however, his understanding of the findings changed dramatically between the two.

The first interpretation, a monograph entitled *The Galapagos Finches (Geospizinae), A Study in Variation*, was published by the California Academy of Sciences. David completed work on the manuscript during his last year at Dartington Hall School and submitted it to the Academy in May of 1940. The monograph was founded not only on the 4 months of field work in the Galapagos but also, and more importantly, on Lack's wing and beak measurements of finch specimens. Based primarily on these measurements, Lack made some modifications to the taxonomy of the finches—modifications that have essentially stood the test of time—and discussed the process of speciation within the group. But as the title of the paper suggests, he focused much of his attention on the tremendous variation in the finches, both within and among species.

Some of the species that are widespread in the archipelago show striking differences among the islands, particularly in beak measurements. Consistent with the then-dominant thinking about sub-specific differences in other bird species, Lack concluded that these differences were nonadaptive and were caused primarily by genetic drift or the so-called founder effect. He wrote, "There is no evidence in favor of the supposition that the differences between the island forms of each species of Geospizinae are of adaptive significance."[14]

He went on to note that Darwin had also been perplexed by the inter-island variation within species, quoting from the sixth edition of *The Origin of Species:*

> This long appeared to me a great difficulty; but it arises in chief part from the deeply-seated error of considering the physical conditions of a country as the most important; whereas it cannot be disputed that the nature of the other species with which each has to compete, is at least as important, and generally a far more important element of success. . . . [W]hen in former times an immigrant settled on one of the islands, or when it subsequently spread from one to another, it would undoubtedly be exposed to different conditions in the different islands, for it would have to compete with a different set of organisms. . . . If then it varied, natural selection would probably favour different varieties in the different islands.[15]
>
> Lack's conclusion regarding Darwin's hypothesis on inter-island variation was blunt: "There is no evidence in favor of Darwin's suggestion. In fact, there is no evidence whatever, in any of the island forms of Geospizinae, that their differences have adaptive significance."[16]

Lack found that the principal differences between closely related finch species concerned bill size and shape, and he came to a similar conclusion about these differences: "There is no evidence that in closely related species the bill differences are related in any way to differences in food, feeding habits, or other differences in the ways of life of the species."[17] He inferred instead that the bill differences among species serve as reproductive isolating mechanisms: "This strongly suggests that the different species recognize each other primarily by their bill differences, and so keep segregated."[18]

Finally, based on both his field observations and his specimen measurements, Lack rejected a suggestion made by some earlier workers that the closely related species were in fact "hybrid swarms." However, he did conclude that the ground-finch populations on two of the small islands, Crossman and Daphne Major, were composed of hybrids between small and medium ground-finches.

As a result of delays caused by World War II, this paper was not published until 1945. By that time, however, Lack's views had completely changed.

During his wartime service with the Army Operational Research Group (see Chapter 2), Lack completed work on *The Life of the Robin* (in 1942) and began work on another book on the Galapagos finches, this one for sixth form students. A dawning appreciation of the implications of Gause's principle of competitive exclusion led him to a rethinking of his earlier conclusions about within-species and among-species variation in the finches, and the new book evolved into a complete reanalysis of his Galapagos material. He completed the manuscript in 1944, and after a rejection from H. F. & G. Witherby, Cambridge University Press agreed to publish *Darwin's Finches*, which was released in 1947.

Although it is not as frequently cited as *Ecological Adaptations for Breeding in Birds, The Natural Regulation of Animal Numbers*, or even *Population Studies of Birds, Darwin's Finches* is arguably Lack's most significant book. In this volume, Lack essentially reversed his earlier conclusions: "Further consideration has led me to realize that the absence of adaptive differences is only apparent, and that in fact closely related species differ from each other in ways which play an extremely important part in determining their survival."[19]

Although he continued to assert that some of the variation in the finches appeared to be nonadaptive, his overarching thesis now recognized that most of the differences within and among species were the product of natural selection. He accepted Mayr's proposition that speciation in birds begins in geographic isolation but asserted that completion of the process requires not only the development of reproductive isolation on re-contact but also ecological isolation of the groups. He identified three mechanisms by which successful speciation had occurred among the Galapagos finches:

1. One of the incipient species was better adapted for a part of their joint geographic range, while the other was better adapted to another part, resulting in geographic separation,
2. One was better adapted to one type of habitat within their range, while the other was better adapted to another habitat, resulting in habitat segregation, and
3. One was better adapted for exploiting one type of food, while the other was better adapted for exploiting another type, resulting in co-existence in the same habitat by partitioning the food niche.

As a salient example, he interpreted the differences in beak sizes of the three ground-finches co-inhabiting several of the islands to be a result of their adaptations for taking different sizes of seeds.

Another reinterpretation involved the status of the ground-finches inhabiting the small islands of Crossman and Daphne Major. David now concluded that their intermediate beak sizes were not caused by hybridization but by what would later be termed "competitive release." He found that the population inhabiting Cross-man consisted of small ground-finches with larger than normal beaks, whereas the population on Daphne Major consisted of medium ground-finches with smaller than normal beaks. The figure illustrating this conclusion became one of the most frequently reproduced figures in ecology or evolutionary biology, appearing in the majority of general biology textbooks published during the last 50 years (Figure 9).

Lack's first presentation of these new views on the significance of the variations in the Galapagos finches had been made in a symposium on the ecology of closely related species at the annual meeting of the British Ecological Society held at Burl-ington House, London, on March 21, 1944. The aim of the symposium was to eval-uate the significance of Gause's hypothesis for the field of ecology, and David Lack was the first speaker. He, Charles Elton, and George Varley (a colleague of Lack's with the Army Operational Research Group who later became Hope Professor of Entomology at Oxford) maintained that the competitive exclusion principle had significant ramifications for the understanding of community structure, whereas the other symposium participants suggested that it did not. Lack recalled later, "For almost the only time in our lives, Charles Elton and I were on the same side, he through his studies of animal communities."[20]

The Significance of Clutch-Size

In 1943, David was appointed biological assistant to the editor of *Ibis*. In that ca-pacity, he received a copy of Reg Moreau's submission describing the latitudinal gra-dient in clutch size in African birds, which was published by the journal the following year. Lack recalled later that, on reading Moreau's manuscript, "it came to me in a flash that the clutch-size of nidicolous birds must have been evolved in relation to the number of young which they can feed and raise."[21] This idea ran counter to the prevailing view among biologists at the time, which suggested that mortality rates drive reproductive rates. This thinking is reflected in a quotation from the 1946 edi-tion of Pettingill's widely used ornithology text:

Species nesting in southern, tropical latitudes have a smaller clutch than closely allied species nesting in northern, temperate latitudes. . . . [A] northern climate exposes birds to greater risks, thus making necessary the laying of more eggs in order to perpetuate the species.[22]

Arguing on the basis of the recently articulated neo-Darwinian Synthesis, Lack maintained that natural selection would always favor individuals that maximize their reproductive output. Individuals that reproduce at lower than their maximum

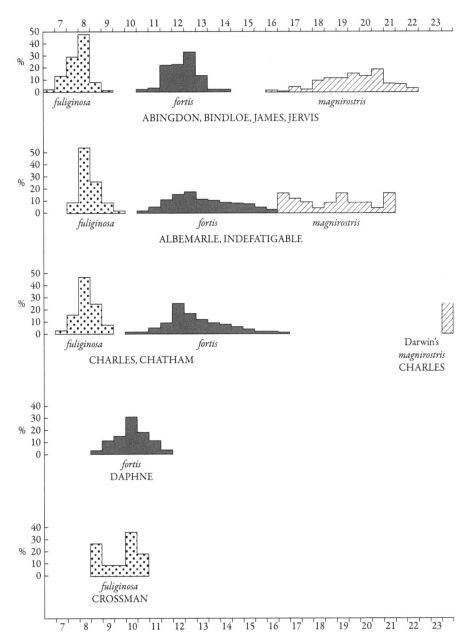

FIGURE 9: Figure from *Darwin's Finches* depicting beak depths of ground-finches from several of the Galapagos islands: *Geospiza fuliginosa*, small ground-finch; *G. fortis*, medium ground-finch; *G. magnirostris*, large ground-finch. Note species designations of finches on Crossman and Daphne Major, which Lack had previously identified as hybrid swarms between small and medium ground-finches. (From D. Lack, *Darwin's Finches*, Figure 17. Reproduced by the courtesy and with the permission of Cambridge University Press.)

rate would be rapidly out-competed by individuals that produce the most offspring possible. Lack concluded that the clutch size of birds that feed their young in the nest would be determined by the amount of food and the ability of the parents to adequately nourish their young. A succinct statement of Lack's hypothesis in his own words is found in *The Natural Regulation of Animal Numbers:* "In most birds, clutch size has been evolved through natural selection to correspond with the largest number of young for which the parents can on the average find enough food."[23]

The first paper in which Lack articulated this view was entitled "Clutch and Brood Size in the Robin" (*British Birds*, 1946), but the major paper, "The Significance of Clutch-Size, I–III," was published in 1947/48. This remains by far the most frequently cited paper ever published in *Ibis*.

Considering Lack's clutch size hypothesis in a broad context, it represented the first optimization model in ecology or animal behavior. As such, it was an innovation that contributed substantially to the transformation of the field of ecology from a primarily descriptive discipline to a more predictive and experimental science. Specifically, the principal method of testing Lack's hypothesis was brood-size manipulation, in which newly hatched young are transferred from some nests to others to create both enlarged and reduced brood sizes. Lack's hypothesis predicted that the normal clutch size in an area would produce more surviving young on average than either enlarged or reduced brood sizes. Many such studies were performed on numerous species during the next two decades, with equivocal results. In about half of the studies, the normal brood size proved to be the most productive, but in the other half, enlarged broods were more productive than the normal brood size.

Numerous modifications of Lack's hypothesis have been suggested to explain the equivocal results of the manipulation experiments, some by Lack himself, but with these modifications, his hypothesis continues to serve as the basis for research in life history strategies. Today, the working hypothesis might be summarized by stating that the reproductive strategy of a species, of which clutch or brood size is a key component, has evolved to maximize the lifetime reproductive potential of the individual. This hypothesis has been used to explain the diversity of reproductive behaviors in not only birds but a wide range of taxa, a fact that helps explain the many citations of Lack's 1947 *Ibis* paper.

FATHER OF EVOLUTIONARY ECOLOGY

Although there are several important progenitors in the field, the two 1947 publications by David Lack had such salient effects on the discipline of ecology that I believe it is fair to recognize him as the father of evolutionary ecology. No less an authority than Ernst Mayr credited Lack with identifying the necessity of ecological isolation in the process of speciation, an appreciation that no doubt contributed to the invitation extended to David to participate in the Princeton Conference in January 1947 (see Chapter 5).

Two of the other pioneers in the field, the Americans Gordon Orians and Robert MacArthur, each spent a year working with Lack at the Edward Grey Institute of Field Ornithology. Orians came to Oxford in 1954 on a Fulbright fellowship after completing his undergraduate studies at the University of Wisconsin (see Chapter 5), and MacArthur came as a postdoctoral fellow in 1957 after completing his Ph.D. at Yale under G. Evelyn Hutchinson (see Chapter 8). Ironically, and perhaps not coincidentally, Hutchinson, another pioneering 20th-century ecologist, had also attended Gresham's School (see Chapter 13).

The History of the Robins gave to children a human tale which they could enjoy and from which they could profit, it was written with grace and charm, and it included much interesting description of country life, many pleasing digressions, and sound moral common-sense. The last is, of course, appreciated by children. It is no surprise that the book was heard with enthusiasm when read aloud at bedtime to the supposedly amoral offspring of left-wing Liberals at Dartington Hall in the nineteen-thirties; after all, their spiritual ancestor, young Mary Godwin, had approved it before them.

—*Robin Redbreast*, p 133

4

Robin Redbreast

∽

DAVID LACK'S LIFELONG love of literature and of the robin came together in his charming fourth book, *Robin Redbreast*, which was published by Clarendon Press in 1950. This little book also reflected Lack's interest in communicating with a general audience. It was organized topically, and somewhat chronologically, and described the varied contexts in which the robin figures in the history of British life and literature. It might be described as an anthology, but Lack provided a rich narrative context for the many poems and stories that were reproduced in each of the 12 chapters.

The first chapter, "The Winter Friend," was based on the tameness and friendliness of robins in winter, when they often come to be fed on the stoop or even enter the house through an open window to feed. The recognition of this behavior predated Chaucer, who characterized the robin as "tame ruddock"—*rudduc* being its Anglo-Saxon name. As in subsequent chapters, Lack traced the history of references to the robin as a "winter friend" in Britain, where, as he stated, "In winter, the robin finds a home with people of all types. Indeed, it enters not only the houses of men, but the house of God."[1] The latter observation refers to a 17th-century Latin poem about a robin that regularly entered Canterbury Cathedral during services.

The second and third chapters, "The Death of the Robin" and "The Covering of Dead Bodies," dealt with subjects that are also rooted in ancient traditional literature. The former began with an anonymous poem, one of the best known nursery rhymes in the English language: *Who Killed Cock Robin?* Although the imagery of many nursery rhymes makes allusions to contemporaneous social or political events, Lack concluded that *Who Killed Cock Robin?* either had no such referents or, if it had them at the time of its origins, they were lost in prehistory. Other literary allusions indicated to him that the poem predates Shakespeare.

The third chapter described an ancient belief that robins cover the faces or the entire bodies of dead persons in the forest with leaves or moss. *The Children in the Wood*, an ancient ballad that Lack dated back to at least the late 16th century, is the possible source of the belief. The tale describes the tragic fate of two children who were committed by a dying Norfolk gentleman into the care of a brother, only to die of exposure after being abandoned in the forest by men hired by their uncle to kill them. A late 17th century version of the ballad, reproduced in its entirety in *Robin Redbreast*, reads, in part, as follows:

> Thus wandered these poor innocents,
>   Till death did end their grief;
> In one another's arms they died,
>   As wanting due relief:
> No burial this pretty pair
>   From any man receives,
> Till robin redbreast piously
>   Did cover them with leaves.[2]

The remainder of the ballad recounts the punishments visited by God on the wicked uncle, and it ends with a moral for executors caring for orphaned children. Lack then proceeded to discuss several other literary references to the covering of the dead by robins.

Subsequent chapters covered literary references to the song of the robin, its red breast, its mating and nesting behaviors, and the robin in politics, among other topics. The final two chapters, "Robins in the Nursery" and "A Happy Christmas," described the place of the robin in two relatively recent aspects of British life and literature, books written specifically for children and the exchanging of Christmas cards.

The epigraph at the beginning of this chapter refers to a book by Sarah Trimmer that was first published in 1786 under the title *Fabulous Histories Designed for the Instruction of Young Children Respecting their Treatment of Animals* (little wonder that the title was later shortened to *The History of the Robins*). Trimmer was one of the pioneers in the emergence of a new genre, literature written specifically for children (a genre celebrated more than 200 years later as part of the opening ceremony for the London Olympic Games in 2012). Lack's assessment is that *The History of the Robins* was probably the preeminent example of such literature until the publication of *Alice's Adventures in Wonderland* in 1865. He writes:

In one form or another, *The History of the Robins* was probably read to most of the great figures of Victorian England, in the days before their greatness. It was evidently a great favourite of the Great Queen herself, as Prince Waldemar received a copy of the Harrison Weir edition inscribed 'given to Waldemar by his dear Grandmamma of England, Xmas 1874.'[3]

Mary (Godwin) Shelley was evidently fond of it as well, because her nickname as a child was "Peeksie," one of the four robin children in the story.

The association of the robin with Christmas cards dates to the advent of the practice of sending Christmas cards around 1860. The earliest English Christmas cards featured the robin, often with an envelope in its beak. Lack suggested that the robin, with its red breast, was representative of the postman, who wore a red waistcoat in the early days of the British postal service. Other possible associations of the robin with Christmas included the fact that it remains in Britain all winter, that it sings in the wintertime, and that it is extremely friendly during the winter, as described in the first chapter of *Robin Redbreast*. Whatever the reason for the prominent place of the robin on English Christmas cards, it is an association that continues to the present day.

More than 50 years after the publication of *Robin Redbreast*, Andrew Lack, David's son, published an updated and expanded version entitled *Redbreast: The Robin in Life and Literature* (SMN Books, 2008). He retained the basic structure of *Robin Redbreast* and likewise organized the material into 12 chapters, several with the same names as those in his father's book. He added numerous new quotations, both from more recent literature and from older sources that were either omitted or overlooked by his father. The new book sold well from the beginning, unlike his father's, which David had envisioned would sell very well but instead had disappointing sales. David blamed this in part on the publisher, which delayed publication until spring instead of releasing the book in time for Christmas. The difficult economic times in England in 1950, with postwar rationing still a part of everyday life, may also have depressed sales.

Robin Redbreast reflected not only Lack's love of literature but also his interest in communicating with the general public, an interest that dated at least to his student days at Cambridge. Two of his other books, *The Life of the Robin* and *Swifts in a Tower*, also demonstrated this interest. Although both were composed with scientific rigor, they were written in a style that is accessible to a general audience. However, like many other scientists, David was ambivalent about the time spent in writing for popular science magazines and other public media, believing that it was a distraction from his "real work." The remainder of this chapter is devoted to describing some of these public efforts.

NEWSPAPERS AND MAGAZINES

The two articles that Lack wrote for the *Magdalene College Magazine* describing his expeditions to St. Kilda and Bear Island have already been mentioned (see Chapter 1). He also wrote a lengthy article for *The Illustrated London News*, the world's first illustrated weekly newspaper. The two-page article, entitled "The Effect of the Exodus from St. Kilda upon the Island's Flora and Fauna: Interesting Changes Observed During a Recent Visit," was published on December 26, 1931, four months after the return of the expedition from the island. After briefly describing the evacuation from St. Kilda of the entire population and their livestock, Lack identified the significance of the observations:

> There was thus created a problem of extreme scientific interest. How would the fauna and flora of this small and remote island, which is only some two-and-a-half square miles in area, be altered by the exodus.[4]

He proceeded to summarize some of the changes that were already evident only 1 year after the evacuation, including sharp declines in the St. Kilda house-mouse, which he described as "extremely rare" and probably doomed to extinction, and in three bird species (hooded crow, tree sparrow, and starling). All of these species were at least partially dependent on humans or their livestock. On the other hand, several seabird species that nest on St. Kilda were flourishing.

Lack included some interesting anecdotes in the article, which was also accompanied, not surprisingly given the name of the publication, by several photographs.[5] One of the anecdotes described a meal that the members of the expedition prepared, "young Fulmar fried in its own fat," which was a staple of the former St. Kildans' diet. Some of the party thought it delicious: "It tasted like chicken saturated in olive oil." Another anecdote related that St. Kilda was one of the last strongholds of the extinct Great Auk, and that on one of the islands "one of the last of these birds is said to have been killed (as a witch!) in about 1840." The photographs included one of the St. Kilda graveyard, surrounded by a circular stone fence, and another of a "cleit," a circular stone building with a turf roof that was used for storing fodder and fuel for the winter. The article thus combined human interest aspects with the results of a scientific expedition to promote a conservation theme.

Lack also wrote an article entitled "Our Feathered Friends—How to Take Them" for *The Kodak Magazine* while he was still a student. Following the publication of *The Life of the Robin*, a condensed version was published in the November 1943 issue of *World Digest*, under the title "Robins Must Advertise."

After he became Director of the Edward Grey Institute of Field Ornithology (EGI) in Oxford (see Chapter 5), Lack continued to contribute to newspapers. In

November 1953, *The New Statesman and Nation* published an article entitled "Bird Art." The following August, "Swifts in an Oxford Tower," co-authored by David and his wife, Elizabeth, was published by *The Times*. In June 1958, on the centenary of Darwin's and Wallace's presentations of their independently conceived ideas of evolution by natural selection to the Linnaean Society of London, *The Glasgow Herald* published David's "Darwin's Theory of Evolution: Natural Selection."

POPULAR SCIENTIFIC PERIODICALS

Beginning after the publication of *Darwin's Finches* in 1947 (see Chapter 3), David contributed regularly to popular science magazines. The first such contribution was an article in *The New Scientist*. It was actually an examination David had written to provide a lighthearted evening for those attending the third annual Student Conference in Bird Biology (see Chapter 5), hosted by the EGI. It was entitled "An Ornithological Examination Paper," and David reported that an Oxford undergraduate scored highest on the examination, about 50 percent, followed closely by a Cambridge undergraduate. (The "Oxbridge" rivalry extended even to lighthearted competition.) The examination was reprinted in Lack's later book, *Enjoying Ornithology* (see Chapter 8).

Subsequent articles by Lack appearing in *The New Scientist* included "Robins for Christmas" in 1960, "Are Bird Populations Regulated?" in 1966, and "Of Birds and Men" in 1969. The last article constituted Lack's entry into the public fray over the biological basis for human aggression, which was instigated by the publication of Robert Ardrey's *The Territorial Imperative: A Personal Inquiry into the Animal Origins of Property and Nations* (Atheneum) and Konrad Lorenz's *On Aggression* (Harcourt, Brace & World), both in 1966. Two years later, the British American anthropologist, Ashley Montagu, edited a strident rebuttal of Ardrey's and Lorenz's views (*Man and Aggression*, Oxford University Press). The controversial Vietnam War raging at the time sparked added public interest in the issue of human aggression. In "Of Birds and Men," Lack described the contexts and ecological consequences of ritualized conflicts in birds, in which actual aggression ensues only rarely, and concluded:

> We cannot seriously hold with Ardrey . . . that 'we act as we do for reasons of our evolutionary past, not our cultural present.' Certainly our evolutionary past is involved, but so are our culture and our reason. If we are to end human warfare, we should seek first to change not Man's aggressive drive, but the Malthusian compulsion. But even that would not suffice, because wars are fought not merely for the necessities of life, but for riches. . . . Certainly we can learn from

the birds, but just as in times past, the devil quoted scripture for his purpose, it is all too easy to draw wrong conclusions through selecting only part of the evidence from the book of Nature.[6]

David thus expressed not only his views on the question of human aggression but also his opinion on another contemporary subject of public concern, that of unchecked growth in the human population.

David contributed two articles to *Scientific American*. The first, "Darwin's Finches," was published in April 1953 and was a summary of his findings published in his book of the same name (see Chapter 3). The article was reprinted in *Scientific American*'s compilation of papers on avian biology, *Birds* (W. H. Freeman, 1980). The second article, "The Home Life of the Swift," was co-authored by David and Elizabeth and was published in 1954. It was a summary of their joint work on the common swift and its many unique adaptations for life on the wing (see Chapter 6).

In 1963, Lack published a short article in *American Scientist* highlighting a rather arcane but immensely influential piece of the history of evolution dating from Darwin's visit to the Galapagos Islands in 1835. The article, entitled "Mr. Lawson of Charles," recounted an enlightening exchange between Darwin and the Islands' Vice-Governor, Mr. Lawson. Lawson informed Darwin that the tortoises from island to island were distinctively different and that he could accurately identify from which island a tortoise had been captured when it was brought to Charles. Until then, Darwin had mixed some of his specimens from the two islands he had already visited, but from that time on he kept the specimens from different islands separated, and his analysis of variation among the birds, reptiles, and plants of the various islands contributed significantly to his dawning awareness of the process of evolution. Lack concluded that Darwin may not have recognized the variation among the species occupying different islands had it not been for his encounter with Lawson: "Whether Darwin would have recognized the fact for himself may be doubted, as his visit was short and the differences between the tortoises are not striking."[7]

RADIO BROADCASTS

Radio was another medium for communicating his science with a general audience. In May 1953, David presented a program, entitled "How Birds Behave II: The Robin," in the BBC *Nature Parliament* series hosted by James Fisher. David also presented "Screening the Migrants" on BBC Radio in October 1958; this program featured the work being done at the EGI on bird migration (see Chapter 8).

After the publication of *Evolutionary Theory and Christian Belief, the Unresolved Conflict* in 1957 (see Chapter 7), David appeared twice on BBC Radio discussing issues in philosophy, science, and religion. The first broadcast, in July 1958, featured a debate between Lack and Dr. David Newth, an embryologist at University College, London; it was moderated by Peter Medawar. The second presentation was broadcast in October 1959 as part of a series on religion and philosophy.

PUBLIC PRESENTATIONS

Again beginning while he was still at Cambridge, David made numerous presentations to clubs and other organizations. After the expedition to St. Kilda in 1931, he returned to Gresham's School in February 1932 to make a presentation to the Natural History Society; he titled it, "Life on Deserted St. Kilda."

Lack also made several presentations to the Oxford Ornithological Society (OOS) beginning even before he became Director of the EGI. In January 1945, he gave a talk entitled "Control of Bird Populations." At a joint meeting of the British Trust for Ornithology and the OOS in October 1948, David and Peter Hartley presented "A Discussion of the Significance of Territory." At the time, Harley was Senior Research Officer in the EGI.

ENCYCLOPEDIA BRITANNICA

David was asked to write an article on biological populations for a periodic revision of the 14th edition of *Encyclopedia Britannica*. He wrote a section entitled "Population, Biological" during 1970–1971 that appeared in several subsequent revisions of the encyclopedia.[8] The article dealt primarily with animal populations but also described some of the difficulties of working with plant populations, particularly how to identify individuals in a clonal species. The article, which covered almost 15 pages in the 1980 edition, was divided into four sections: (1) the study of populations, (2) the characteristics of biological populations, (3) changes in population characteristics, and (4) interactions of populations.

TELEVISION

Although David did not appear on television himself, a part of his life, including some of his science, was presented in a fictionalized docudrama on BBC television in 1999. The show, *Wings of Angels*, was historically inaccurate in a number of key details, but it did focus on two major struggles in David's life, one scientific and one personal. The scientific conflict involved his difficulty in interpreting his findings on the finches of the Galapagos Islands (see Chapter 3), and the personal struggle centered on his conversion to Christianity from agnosticism in the context of his developing romantic relationship with his field assistant and future wife, Elizabeth Silva (see Chapters 6 and 7). A further scientific element in the drama, which provided its name, concerned the discovery during World War II, by George Varley and Lack, of the fact that the moving images on radar screens often referred to as "angel's wings" were actually the images of flying birds (see Chapter 2). One of Lack's sons recalled that he had the "surreal experience of watching the, remarkably accurate, fictional version of my mother with my actual mother sitting beside me."[9]

If an animal is introduced to a new and favourable area, it at first increases rapidly, but it is soon checked, and thereafter its numbers, like those of other animals, fluctuate between limits that are extremely restricted compared with what is theoretically possible. It follows that natural populations are in some way regulated, and that the controlling factors act more severely when numbers are high than when they are low.

— *The Natural Regulation of Animal Numbers*, p 275

5

The Natural Regulation of Animal Numbers

ON THE EVE of the 50th anniversary of the publication of *The Natural Regulation of Animal Numbers*, one of David Lack's doctoral students, Ian Newton, published an article entitled "Population Regulation in Birds: Is There Anything New Since David Lack?"[1] Although Newton's conclusion was that indeed there had been significant advances in the understanding of population regulation in birds, the fact that he could publish a paper with that title in 2003 attests to the enormous impact of *The Natural Regulation of Animal Numbers* on the field of avian population biology. In fact, its influence on population ecology extended far beyond ornithology, and it is Lack's most frequently cited work, with Google Scholar listing more than 3400 citations since its publication.

The Natural Regulation of Animal Numbers also represented a significant departure from Lack's previous work in two major ways: (1) it focused on a general topic, the regulation of animal populations, and (2) it included a survey and analysis of the literature on population regulation in all groups of animals, not just birds. In many respects, it is the most ambitious of his books in its breadth and scope. Similar synoptic books published later included *Population Studies of Birds* (1966), *Ecological Adaptations for Breeding in Birds* (1968), and *Ecological Isolation in Birds* (1971).

Lack's principal argument in *The Natural Regulation of Animal Numbers* is summarized in the epigraph at the beginning of this chapter. In the introductory first chapter, he introduced two ideas that were central to his arguments throughout the book but which he said had been difficult for biologists to appreciate. The first was that "the reproductive rate of each species is a result of natural selection, and is not,

75

as often supposed, adjusted to the mortality rate of the species."[2] The second was that "the critical mortality factors are density-dependent, hence climate *per se* cannot be the primary factor controlling numbers."[3] The first of these ideas was derived from Lack's own work on clutch size in birds, described in Chapter 3. The second was based on a concept of density dependence developed by the Australian entomologist, A. J. Nicholson, in the early 1930s. Density-dependent factors result in lower reproductive rates and/or higher mortality rates as population density increases, whereas density-independent mortality factors (such as a severe weather event) tend to result in the death of a fixed proportion of the population irrespective of density.

The first chapter after the introduction described the results of several long-term censuses of animal populations, principally bird populations, that displayed, according to Lack, a relative stability compared with their potential for exponential growth. He concluded that this situation implies that the populations are regulated in some way and, further, that the factors responsible for such regulation must be density dependent.

Several subsequent chapters discussed the evidence for the assertion that species of birds and other animals are selected to reproduce at the maximum rate possible. One novel extension of Lack's earlier work on clutch size determination in birds was included in these chapters—a concise statement of his "brood reduction hypothesis" as the adaptive explanation for asynchronous hatching in birds:

> Many species of birds start incubating only when the clutch is complete, but hawks, owls, storks, crows, and various others start when the first egg has been laid. As a result, the earlier eggs may hatch several days before the later ones. . . . [W]hen food is short the smallest chick gets none, and quickly dies. If food remains scarce, the next smallest chick may also die, but there may be enough food to raise some of the young. On the other hand, if all of the chicks had hatched on the same day and been of the same size, the food might have been divided equally between them, and all might have died.[4]

The adaptive significance of hatching asynchrony is that the brood size can be efficiently adjusted in species for which the nestling food supply or feeding conditions vary from year to year. The hypothesis, which had emerged from Lack's studies of one such species, the common swift, was implicit in an earlier paper on the breeding biology of the species that he co-authored with Elizabeth Lack.[5]

A chapter dealing with the effect of increased density on reproductive rates concluded that the effects, although present in some species, were too small to bring about the necessary reductions in populations at high densities. Lack therefore concluded that density-dependent mortality factors must be the primary regulating factors.

He identified three factors that he considered to be the likeliest means of effecting the necessary density-dependent mortality in populations: (1) predation, (2) parasitism and disease, and (3) competition among members of a population for limited resources such as food or living space. He devoted a chapter to discussion of the evidence for each of these factors in the regulation of bird populations, and his principal conclusion was that food limitation is the most common factor in the regulation of numbers of birds. Although he described numerous studies showing that predation and parasitism/disease often cause significant mortality in birds, he suggested that predation may actually increase population numbers by selectively removing young or weak individuals, thereby increasing food resources for the healthy members of the population, and that individuals succumbing to parasitism or disease may be susceptible because of malnutrition. In short, Lack concluded that food availability is the primary regulating factor in most bird populations.

In other chapters, Lack discussed the evidence from other animal groups, particularly mammals and insects. He found that this evidence also tended to support his general conclusion for bird populations, that food is the primary limiting resource.

One chapter dealt with the special case of cyclic populations—populations that show regular patterns of dramatic change in density, with peaks followed by sharp decreases in numbers. Many of these are populations of arctic or subarctic mammals, with the best known case being that of lemmings, which had long been thought to commit mass suicide by plunging themselves into the sea at times of very high density. Many of these cycles showed periodicities of approximately 4 or 10 years, and they involved several rodent species, including lemmings, voles, and hares, as well as their major predators.

Besides the intrinsic interest in a phenomenon involving such dramatic and regular changes in population densities, two other dimensions enhanced the story even more. First, early mathematical models of predator-prey interactions, developed in the 1920s and 1930s by A. J. Lotka and Vito Volterra, predicted that such cyclic patterns could be caused by the interaction between predator and prey populations. As prey populations increase dramatically in numbers, predator populations increase due to increased availability of prey. Eventually predator numbers increase to such an extent that the prey populations decline rapidly, followed by a similar decline in the predator populations as prey becomes scarce. When the predator populations reach their nadir, prey populations can again increase, initiating another cycle. Although various experimental attempts to verify the predictions of the models had failed, the idea that predator-prey interactions could lead to intrinsic cycles in both populations remained an intriguing one. Second, Lack's closest neighbor, Charles Elton, and his colleagues at the Bureau of Animal Population (BAP) had been pioneers in the study of such cycles and were continuing to work on the problem. In fact, at the time that Lack was writing *The Natural Regulation of Animal Numbers*, Dennis Chitty and his D.Phil. student John Clarke were working on the hypothesis that the sharp declines in rodent populations were caused by internal physiological responses to crowding, and not by increased predation.

For Lack, the fluctuations in predator populations clearly tracked changes in the numbers of prey, consistent with his general conclusion that most populations are food-limited. The more difficult question concerned identification of the underlying cause or causes for the cyclic fluctuations in the rodent populations. Using a form of argument that became something of a pattern in later works, he identified alternative hypotheses for explaining the phenomenon and then examined the evidence to draw his conclusion. In this case, he identified two potential explanations:

> Omitting from consideration the rather extensive lunatic fringe, two main types of explanation have been put forward to explain the rodent cycles, one involving an extrinsic cause, mainly climate, and the other an intrinsic cause, usually a predator-prey interaction.[6]

Lack dismissed climate as an underlying cause, noting that it was not density dependent, which he deemed necessary, and that there were no climatic patterns that tracked the rodent population cycles. He also dismissed predator-prey interaction as incapable of causing the declines, noting that the number of predators was in most cases insufficient to cause the rate of mortality required to bring about the observed decline. He suggested instead that the declines must be initiated either by food shortage or by crowding per se (involving some mechanism such as those being investigated by Chitty and Clarke). He dismissed the latter possibility, however, based on the facts that the declines were initiated at widely differing densities in a species and that the declines persisted for two or three generations despite sharply reduced densities. He summarized his conclusions regarding the factors controlling cyclic rodent (and grouse) populations as follows:

> In conclusion, the view tentatively advanced in this chapter is that cyclic declines are due: (i) to food shortage (with perhaps secondary disease) in the dominant rodents, the birds of prey, and the fur-bearing carnivores, and (ii) to predation in the gallinaceous birds and perhaps in the scarce rodents, after the rodent-predators have switched to them following the decrease of the dominant rodents.[7]

Thus, Lack's explanation of cyclic populations coincided with his broader conclusion that the populations of most species are regulated by density-dependent food limitation. (In actuality, the regulation of rodent and other population cycles is still an active area of research.)

The Natural Regulation of Animal Numbers was favorably reviewed by a number of people writing in both scientific journals and nonscientific publications. J. B. S. Haldane, in a special review in *Ibis*, suggested that the book would be indispensable for any student of animal ecology, and stated: "Just because it is so good, I am going to criticize it severely, on the principle 'You only have I known among all the families of the earth, therefore I will punish you for all your iniquities.'"[8] Haldane identified three areas of concern: (1) lack of mathematics (including failure to properly consider mathematical models and absence of appropriate descriptive and inferential statistics), (2) omission or oversight of some relevant papers from the nonavian literature, and (3) failure to recognize the implications of the differences between "scramble" and "contest" forms of intraspecific competition for food or other resources. The last criticism is particularly interesting, because it presages a heated debate that would embroil Lack for the next decade, ultimately resulting in the publication of another book (see Chapter 9):

> Any animal with a social life must be adapted for behaviour based on a comparatively stable population density. So the evolution from scramble to contest is a pre-requisite for social life. But many birds have taken a further step, from lethal contest to ritualized contest. This is the basis of the territorial system, and I suggest that this system is largely responsible for the stability of bird populations. Lack contests this view on page 260, but I am not sure that he has met all the arguments for it.[9]

Twelve years later, Lack's failure to appreciate the possible role of territorial behavior in regulating bird populations would contribute to the loss of a very talented graduate student (see Chapter 10).

The challenge to Lack's conclusions that led to the debate came primarily from V. C. Wynne-Edwards, who also reviewed *The Natural Regulation of Animal Numbers* in *Discovery*. While praising many features of the book, Wynne-Edwards expressed reservations about some of Lack's conclusions, and this formed the basis of their lengthy dispute (see Chapter 9). In particular, he disagreed with Lack's assertion that natural selection always favored individuals that maximized their reproductive effort:

If the contrary to Dr. Lack's view could be accepted, namely that the reproductive rate is just one of a host of physiological adaptations, adjusted to meet the exigencies of the environment, we should be on familiar and more comfortable ground. In fact, new and promising-looking vistas would open up before us.... One of the most important adaptations, we might expect would be the means of controlling their own population, and not having to rely entirely on external agents to do the job.[10]

Wynne-Edwards was to spend much of the next 7 years writing his own *magnum opus* exploring those "promising-looking vistas."

The Natural Regulation of Animal Numbers, which was Lack's first major scientific book based on his work as Director of the Edward Grey Institute of Field Ornithology in Oxford, was republished in hardcover by Oxford University Press in 1967 and in paperback in 1970. It was also translated into Russian.

THE EDWARD GREY INSTITUTE OF FIELD ORNITHOLOGY

In 1932, Edward Grey, Viscount of Fallodon, was Chancellor of Oxford University, having been appointed to that position in 1928. His undergraduate achievements at Oxford would not have suggested that he would someday become the University's Chancellor. He entered Balliol College in 1880 to read classics, was sent down 2 years later "for idleness and ignorance," but returned in 1884 to earn a Third Class degree in jurisprudence. A year later, he was elected as a Liberal Member of Parliament from Northumberland, serving in William Gladstone's Liberal government. He continued to be active in the Liberal Party throughout the next 20 years, during several of which the party was out of power. When the Liberals were returned to power in 1905, Grey was appointed Foreign Secretary by Prime Minister Henry Campbell-Bannerman. He continued in that post under Campbell-Bannerman and H. H. Asquith until December 1916, when David Lloyd-George became Prime Minister. More pertinent to the present account, Grey had a lifelong avocational interest in birds, particularly ducks. One of the best-known photographs of Grey shows him with a robin perched on his hat.

The institute that bears Viscount Grey's name began rather inauspiciously as an outgrowth of the Oxford Ornithological Society. In 1927, Max Nicholson, then an undergraduate at Hertford College reading modern history, and other members of the Society, including its vice-president, Bernard W. Tucker, and its secretary, V. C. Wynne-Edwards, initiated the Oxford Bird Census. This project organized censuses of several bird species using Oxford undergraduates and amateur bird enthusiasts, primarily in the Oxford area, but including a national census of heronries in 1928. The Census also organized trapping and ringing programs, ringing 1650 birds by 1931. The success of these efforts led Nicholson and Tucker, then University Demonstrator in Zoology and Comparative Anatomy, to propose that a permanent center for the study of field ornithology be established in Oxford. Sufficient funds, including grants from the Empire Marketing Board, the Ministry of Agriculture, and private donors, were obtained to hire a Director for the Census, W. B. Alexander. An advisory committee was established with E. S. Goodrich, Linacre Professor and Head of the Department of Zoology and Comparative Anatomy, as chair. Other members included representatives of the Oxford Ornithological Society, the Ministry of Agriculture, and the University Departments of Zoology, Entomology, and Botany and Agriculture.

Wilfrid Backhouse Alexander was born in Croydon, Surrey, in 1885 and educated at King's College, Cambridge. Before becoming Director of the Census, he had spent most of his career in Australia, where he had developed a considerable reputation as a zoologist. From 1920 to 1925, he served as the biologist on the Commonwealth Prickly Pear Board, which was attempting to find a solution to the problem of the devastatingly invasive prickly pear cactus in Australia. In this position, he was the first to bring live prickly pear moths *(Cactoblastis)* from Argentina to Queensland. The success of this program is legendary in the field of biological control.

Under Alexander's leadership, a proposal for the creation of an Institute of Field and Economic Ornithology at Oxford was prepared by the advisory committee in 1932 and submitted to the University. The introduction of the proposal stated that scientific progress was frequently impeded by lack of data:

> At any rate in the case of ornithology this starvation of science is not the result of any deficiency of observers. It is due simply to a deficiency of scientific direction and organisation. The quantity and the quality of observers are adequate, but they would benefit by training and guidance. There is need not only for encouraging work of an individual character but also for developing a tradition of team-work under competent leadership.[11]

The "essentials of corporate observation" were identified in the proposal as (1) concentration of aim, (2) expert direction, and (3) training of observers. After elaboration of the aims of the future institute, the proposal concluded with an appeal for funding:

> It might be said that this is an ill-chosen moment to embark on so large a scheme. . . . There can be no question at the present time of going beyond bare

necessities. It is therefore proposed to limit the appeal to an endowment of £1600 per annum for not less than 5 years, which it is estimated will enable the essentials of the scheme to be developed subject to strict economy.[12]

This proposal led to the founding of the British Trust for Ornithology (BTO) in 1933, which was initially established to provide support for the new institute and its leadership in coordinating census and ringing activities. The wheels of Oxford University grind with exceeding slowness, and it was not until 1938 that the relationship of the fledgling institute with the university was finalized. Chancellor Grey, a supporter of the institute, had died in 1933, and when the university affiliation was finally formalized, it was named the Edward Grey Institute of Field Ornithology (EGI). W. B. Alexander remained as Director throughout this protracted process, and he continued in that post until he retired in 1945.

A NEW DIRECTOR OF THE EGI

The year 1945 may have marked the end of European hostilities, but it also saw something of a battle for the future direction of the EGI after Alexander's retirement. Although a January 1945 report of the Committee on the Future of the Edward Grey Institute of Ornithology to the Hebdomnal Council of Oxford University stated that there was "one good man available for the post," the Oxford Committee for Ornithology sent a letter dated July 4, 1945, to three people—James Fisher, David Lack, and H. N. Southern. The letter stated, in part:

> The appointment will begin from the retirement of the present Director at the beginning of the Michaelmas Term, 1945, and will be in the first instance, for one year on a salary scale of £450–£550, according to age and experience. . . . It is hoped that within a year the University will be able to confirm the continuation of the appointment, in which case the salary will rise to a maximum of £750. It is expected that, in addition to the duties in the Institute, the new Director will be asked to assist in the Honours Course in the Department of Zoology and Comparative Anatomy.[13]

The letter asked that those interested in being considered for the post submit 10 copies of their application by July 12 and that the application include their name, age, nationality, academic qualifications, and a list of publications, as well as the names of two referees. All three responded to the invitation, Fisher and Lack submitting the requested application materials and Southern responding in a curious manner. He wrote,

I should like to make it clear that I only wish to apply for this post in the event of Mr. David Lack's candidature lapsing. Therefore, although I hereby make a formal application to be considered by the Committee, I only wish it to take effect under such circumstances.[14]

At the time, Southern held a post in the BAP. Fisher named Julian Huxley and Max Nicholson as his two referees, whereas Lack named William Thorpe of Cambridge University and N. B. Kinnear, natural history keeper at the British Museum.

The decision-making process of the Oxford Committee for Ornithology is unknown. Max Nicholson, who was at the time very active in the BTO and would serve as its chairman from 1947 to 1949, strongly supported Fisher's candidacy. Fisher, who had achieved Second Class honors in zoology at Magdalen College, Oxford, was also active in the BTO. He had joined the organization in 1936 and had served as its honorary assistant secretary, treasurer, and honorary secretary during the intervening years. He had also been a member of the Advisory Scientific Committee of the BTO and the EGI since 1940 and had organized the EGI's rook investigation during World War II. He had published three books: *Birds as Animals* (Hutchinson, 1939), *Watching Birds* (Penguin, 1940), and *The Birds of Britain* (Collins, 1942).

Julian Huxley, who had served as Lack's unofficial mentor for more than a decade, was apparently Lack's major champion. As I have discussed in earlier chapters, Lack's principal accomplishments to date were as a schoolmaster at Dartington Hall, author of *The Life of the Robin*, and writer of several widely acclaimed scientific papers including those on territory and habitat selection.

The contest was apparently an intense one. I could find no record of the deliberations. A terse telegram to W. B. Alexander, dated July 27, 1945, from Bernard Tucker read, "NORWOOD AND CHAPION AGREE HAS GOODRICH REPLIED = TUCKER."[15] The retiring Head of the Department of Zoology and Comparative Anatomy at Oxford, E. S. Goodrich, was apparently a member of the committee, but it would be fair to surmise that the incoming Head of the Department, Alister Hardy, also had something to say about the outcome. Tucker, by now a Lecturer in Zoology (and soon to be appointed Reader in Ornithology), no doubt had an important say. Charles Elton, Director of the BAP, may also have weighed in in Lack's favor, because he had found Fisher's work habits wanting when he participated in a wartime BAP project and had urged him to seek other employment more suitable to his gifts.[16] In any event, David Lack was selected to succeed Alexander as the second Director of the EGI. As shall be detailed later, Lack's and Nicholson's visions for the EGI and its relationship to the BTO differed dramatically. These differences and the intensity of the contest over the appointment were manifested by the fact that Lack's relations with Nicholson remained strained for the rest of David's life.

David Lack arrived in Oxford in August 1945, and he assumed the directorship of the EGI on Alexander's retirement October 1st. David had been released early from his service with the Army Operational Research Group because he had employment. Two months after assuming the directorship, the EGI moved from its cramped, wartime quarters at 39 Museum Road to a house at 7 Keble Road. The new quarters consisted of five rooms instead of the three on Museum Road, plus basement storage facilities. The library had a much larger room, and Alexander remained at the institute on a volunteer basis as Librarian.

One of David's first priorities was to establish a research focus for the institute. He believed that there were three areas in which the study of birds could make significant scientific contributions: evolutionary biology, ethology, and population ecology. He acknowledged that Ernst Mayr was providing strong leadership in avian evolution and that Niko Tinbergen was providing similar leadership in avian behavior, leaving population ecology of birds as the area in which the EGI could make its mark. Furthermore, his studies of the robin and of Darwin's finches had convinced him that there was much to learn from long-term population studies of birds.

An exchange of letters with Tinbergen in October 1945 makes it apparent that Lack had already chosen ecology. Tinbergen was laying the groundwork for creating a new journal, *Behaviour*. He had first approached Julian Huxley about serving as the English editor for the journal, but Huxley had turned him down and referred him to the newly appointed Director of the EGI. In approaching Lack, Tinbergen wrote, "I could fully understand when you prefer to give your time entirely to your new job, especially when you want to return to ecology."[17] Lack did decline the position and recommended his Cambridge mentor and friend, William Thorpe, who accepted.

Lack's initial plan was to conduct population studies on the robin. Wytham Woods, a large tract of deciduous forest adjacent to the village of Wytham, about 5 km from the center of Oxford, had been bequeathed to the University in 1942 by Raymond ffennell. Charles Elton's BAP was already doing extensive field research in Wytham Woods, and David decided to begin his long-term population studies of the robin there as well. The work that Elton and Lack and their many students did in the following decades in Wytham Woods would ultimately lead John Krebs (Lord

Krebs, of Wytham) to declare: "If there were a Nobel Prize for Ecology, and you could award it to a place rather than a person, Wytham Woods would surely be a prime candidate."[18]

After the 1946 breeding season, however, Lack realized that the difficulty of locating robin nests would make it impossible to obtain sufficient data on the reproductive success of that species. He therefore began to look for another species on which to begin long-term studies.

A trip to Leiden, Holland, in February 1946 at Tinbergen's invitation had already provided Lack with another option for long-term studies in Wytham Woods. While in Holland he had met Hans Kluyver and had become acquainted with the long-term studies on the great tit that had been conducted near Wageningen. Those studies, which had been initiated in 1912 by K. Wolda, had subsequently been continued by Kluyver. The great tit, a secondary cavity nester, readily accepts nest boxes for breeding, negating the problem of search time for nests.

The subject of the symposium to which Tinbergen had invited Lack was "factors determining distribution" in animals, and contributors besides Lack included the Dutch marine ecologist, Jan Verwey, and the Finnish ornithologist, Pontus Palmgren. Tinbergen also invited Julian Huxley and Charles Elton, but both were unable to participate. David presented on the topic, "Competition Between Species as a Limit to Distribution." He also showed his film on the Galapagos finches.

Although they had begun corresponding before the war, David had never met Tinbergen, so Niko wrote that either he or his brother, Luuk, would meet Lack's ship on its arrival: "I now take it as certain that you sail February 14th, and you know somebody with field glasses will pick you up."[19] Because Niko was hosting Palmgren, David's hosts were, successively, Prof. C. J. Van der Klaauw and Luuk. In response to David's inquiry about what he could bring for his hosts, Niko replied:

> Of course, we appreciate very much your kind offer to bring some stuff like tea, coffee etc. with you. The situation here is that we have relatively small, though sufficient rations of the really fundamental foods. What is scarce, are exactly the things that are scarce in your country, too, as for instance meat, fat, leather and clothing. . . . But I know that the Van der Klaauw family could do with a little piece of soap. . . . As regards my brother, he is a smoker who will certainly be very pleased with some genuine English cigarettes.[20]

David remained in Holland for about 2 weeks, and when he returned to Oxford, Niko accompanied him.

After the decision to abandon the robin project, Lack decided to emulate the Dutch tit study, and 110 nest boxes were constructed and put up in Marley Wood, a 60-acre tract on the eastern edge of Wytham Woods. The initiation of the project was funded primarily by a grant from Mrs. J. B. Priestley, wife of the writer and an early benefactor of the EGI during Lack's first years as Director. Observations began there in the spring of 1947. In 1950, 200 additional nest boxes were placed in another part of Wytham Woods.

During the 1946 breeding season, Lack also initiated a pilot study of the common swift in small colonies in two villages near Oxford. His secretary-turned-field-assistant, Elizabeth Silva, did much of the observation on breeding in the swift colonies. Unlike the robin project, the swift project proved productive, and it was expanded and moved in 1948 to the tower at the University Museum on Park Street in Oxford (see Chapter 6).

The development of a new research program for the EGI was accompanied by a significant turnover in the staff of the institute. During World War II, the major research focus had been on bird species with potential economic impacts, particularly on the food supply. Hence, a BTO research team had been assembled to study the rook and the woodpigeon with funding from the Agricultural Research Council. This effort had complemented research being done at the BAP on the control of mice and rats. All the members of this team—M. K. Colquhoun, J. Fisher, R. Harding, Z. King, and I. Werth—left the staff in 1946.

Lack's first appointment, in January 1946, was of a secretary to the Director, Elizabeth Silva. John Gibb, who was later to be one of Lack's first D.Phil. students, joined the staff in July 1946 as a half-time field assistant. In February 1947, Peter H. T. Hartley was added to the staff as a senior research officer. All were to play key roles in the early years of Lack's tenure at the EGI, and one also had a profound effect on his personal life (see Chapter 6). The Agricultural Research Council was now funding the EGI to study the relationship between the control of bird populations and their food supply.

Another early addition to the EGI staff was Reg Moreau. Moreau had retired from the Agricultural Research Center in Tanganyika after an eye infection that threatened his vision. He was initially considered for a position with the new behavior field station at Madingley that William Thorpe was planning to start. However, when plans for the Madingley project bogged down, David, who had kept in contact with Moreau since his visit to Tanganyika in 1934, offered him a half-time appointment as a research officer at the EGI with a "minute honorarium." Reg arrived at the EGI in September 1947, and at the same time he was appointed editor of *Ibis*, also with a small honorarium. Later that year, he and Winnie purchased "an amusing, if tumbledown, cottage, near Oxford."[21] His presence over the next 19 years was to play a crucial role in the developing ethos of the EGI.

Thus, within 2 years, David had assembled a team that would be the heart of the EGI for much of the first 10 years of his directorship. Alexander stayed on as Librarian until his second retirement, in July 1955. Elizabeth and David married in 1949, but she continued as a field assistant until shortly before the birth of their son Peter in January 1952. Gibb eventually completed a D.Phil. in 1953 but continued as a senior research officer with the EGI until 1957. Hartley served as senior research officer until January 1951, when he resigned to become warden of Flatford Field Centre in Suffolk. He would later take Holy Orders and serve as rector at Badingham, also in Suffolk. Moreau worked at the EGI until 1966, when illness prevented him from continuing.

The beginning of Michaelmas term in the fall of 1948 brought the first two D. Phil. students to the EGI. In December 1946, Robert Hinde had attended the first Student Conference in Bird Biology (discussed later) while working on his bachelor's degree in zoology at St. John's College, Cambridge. He obtained a 2-year fellowship to pursue doctoral studies at the EGI. Monica M. Betts completed her undergraduate degree in zoology at University College, London (having attended during World War II, when the college was relocated to North Wales to escape the London blitz). She received support from a 5-year Agricultural Research Council grant to the EGI to study the food of birds. Also in 1948, David Snow, an undergraduate at New College, Oxford, began volunteering as a field assistant in Wytham Woods. The members of the EGI in 1950 are shown in Figure 10.

The daily rhythm of the EGI was established during those early years. The staff would join around a table in the library for coffee at 11 a.m. and for tea at 4 p.m., usually accompanied by a lively discussion. Discussion topics were generally introduced by David and centered on questions that interested him at the time. Ivan Goodbody, who came to the EGI as a student in 1949 with virtually no background in ecology, described them as follows:

> At these tea sessions, David always sat at the head of the table and talked to us. Sometimes it was about a new paper he had read, other times about some new idea he had or a conversation he had with another student. At the time I think I was a bit overawed by David's intellect and the breadth of his knowledge. Those sessions and the conversation they provoked were stimulating and later in life some of the things David had talked about came into my head and influenced my own approach when studying the marine invertebrates of the Caribbean.[22]

Reg Moreau provided a useful foil for David, often expressing an opinion different from that of "the Boss" (an appellation the D.Phil. students and research

FIGURE 10: EGI personnel, 1950. Seated on ground from left: M. T. Myres, Ivan Goodbody. Seated on chairs from left: Peter Hartley, W. B. Alexander, David Lack, Elizabeth Lack, Reg Moreau. Standing from left: R. Hinde, Monica Betts, John Gibb, S. Hewins (secretary), David Snow. (Photograph courtesy of the Edward Grey Institute of Field Ornithology.)

assistants used among themselves when referring to David). Bill Sladen, a D.Phil. student in the early 1950s, described one coffee-time session in which he began discussion of an esoteric botanical subject before David arrived for coffee. David remained uncharacteristically quiet that morning but arrived the next day with a botanical question of his own. Sladen's conclusion was that David exercised a kind of tyrannical proprietorship of the coffee and tea-time discussion topics.

After the EGI moved to the Botanic Garden (discussed later), many took their lunches either in the Alexander Library or in Reg Moreau's spacious office, at least during the winter. In the summer, they often had lunch on the balcony overlooking the Garden with its magnificent beds and borders of perennials.

INAUGURATION OF THE STUDENT CONFERENCES IN BIRD BIOLOGY

In December 1946, David initiated another program at the EGI, the Student Conferences in Bird Biology (which he originally referred to as the Annual Conference on the Biology of Birds for Zoology Students). The first conference featured research papers from both senior scientists and students, including undergraduates. This format reflected Lack's intent to provide students, both undergraduate and graduate, an opportunity to report on their research in an environment that was both sympathetic and constructive. The first conference was attended by 37 persons from 10 universities and focused on the biological aspects of ornithology.

The second Student Conference was held in early January 1948, and they have continued to be held at the beginning of each year since. The theme for the second conference was "The Species Problem in Birds," and 40 students from 14 different colleges and universities participated. The third conference received more applications than the facilities could accommodate, and a dozen applicants had to be turned away. The first international student, from the University of Basel, was among those attending, and in 1950 students from both Switzerland and the Netherlands attended the conference.

The structure of the conferences gradually evolved. They were usually thematic, and Lack began inviting three or four senior scientists to give plenary lectures, while all other talks were by students. The third conference, in January 1949, included the addition of an evening of "light-hearted ornithology." Some of the conferences included trips to the Wildfowl Sanctuary at Slimbridge, Gloucestershire. The themes often reflected Lack's own research interests, such as migration (1952), the population problem (1954), and adaptive radiation (1955). Plenary speakers in the early 1950s included A. J. Cain, Dennis Chitty, Max Nicholson, Peter Scott, Mick Southern, and Niko Tinbergen. At the 1955 conference, Gordon Orians, an American spending a year at the EGI on a Fulbright student fellowship, organized an exhibit on the adaptive radiation of American birds.

Many future British ornithologists made their first presentations at a Student Conference in Bird Biology, and over the years David frequently used the conferences for talent spotting.

THE PRINCETON CONFERENCE

David's third trip to the United States was initiated by a letter from Glenn L. Jepsen of Princeton University, which began,

Dear Dr. Lack:

A series of conferences for eminent scholars and scientists is being planned on various subjects as part of Princeton's celebration of its bicentennial in 1946–47, and this letter is to ask you to participate in the conference on

"Genetics, Paleontology, and Evolution" which is scheduled for January 2, 3, and 4, 1947. A more formal invitation will be extended by the University upon receipt of your favorable reply.

For several years a National Research Council Committee, on Common Problems of Genetics, Paleontology and Systematics, has been working in this general field and the committee is cooperating with Princeton in the Bicentennial conference. The twenty-nine members of the N. R. C. Committee and about thirty other scientists will be invited to be the guests of the University at the Princeton Inn during the conference.[23]

Ernst Mayr, a member of the NRC committee and a prominent participant in the conference, later described its objective as follows:

to examine the rapid changes in evolutionary biology that occurred in the period of the synthesis (from approximately 1935 to 1947), to reconstruct the sequence of events leading to the synthesis, and to identify the factors responsible for the preceding disagreements.[24]

American participants in addition to Jepsen and Mayr included Th. Dobzhansky, H. J. Muller, A. S. Romer, G. G. Simpson, G. L. Stebbins, and S. Wright. Four participants in addition to Lack came from England: the biologists, E. B. Ford, J. B. S. Haldane, and D. M. S. Watson, and the geologist, T. S. Westoll. Princeton covered the trans-Atlantic travel costs of the English participants.

Mayr had actually written Lack in March to inform him that an invitation might be forthcoming for such a conference. After the formal invitation was proffered and accepted, David received offers to visit several American universities during his trip. Invitations came from Charles Kendeigh of the University of Illinois, Lee Dice of Michigan, and Aldo Leopold and John Emlen of Wisconsin. Mayr acted as David's go-between in arranging his American schedule following the conference. The title of David's presentation at the Princeton conference was "The Significance of Ecological Isolation," and the paper also appeared in the publication of the conference proceedings.[25]

After the 3-day conference, David traveled to several eastern and midwestern universities with major research programs in population ecology. In addition to the three institutions already mentioned, these included programs led by W. C. Allee at the University of Chicago, Paul Errington at Iowa State University, and G. Evelyn Hutchinson at Yale. He also visited Margaret Morse Nice in Ohio and Joseph Hickey at the U. S. Fish and Wildlife Service in Patuxent, Maryland, before returning to England.

IN THE GARDEN AT ST. HUGH'S

St. Giles' Square, which lies three blocks north of Carfax, the center of old Oxford, is actually the largest intersection near the heart of Oxford. At the southern end of the square, Cornmarket Street enters from the south and Beaumont Street enters from the west, between the Randolph Hotel and the Ashmolean Museum. Two major arteries, Woodstock Road and Banbury Road, enter the north end of the square. At each end of the square stands a 12th century church— St. Mary Magdalen on its southern border and St. Giles' on the peninsula separating Woodstock and Banbury Roads. The square actually resembles a tree-lined plaza more than an open square. St. John's College stands on its eastern flank, and halfway up the square on the west side is the Eagle and the Child, a pub known the world over as the meeting place of J. R. R. Tolkien, C. S. Lewis, and the Inklings.

St. Hugh's College is located a half-mile up the Banbury Road from the square, at St. Margaret's Road. St. Hugh's was founded in 1886 by Dame Elizabeth Wordsworth (great-niece of the poet) as one of the first women's colleges of Oxford University. The main building faces St. Margaret's and has an expansive garden behind it. A statue of St. Hugh of Lincoln and his swan stands on the stairway, guarding the foyer of the main building.

During World War II, the garden of St. Hugh's was appropriated for use as a head injury hospital for wounded soldiers, led by the eminent neurosurgeon, Sir Hugh Cairns. Several Nissen huts, similar to Quonset huts but constructed of brick walls covered with a rounded tin roof, were built in the garden behind the main building. As one contemporary student described it, "[We could] from our window see the Nissen huts smothering the rose-beds and tennis courts."[26] The head injury hospital ceased operations shortly after the war, vacating the Nissen huts in the St. Hugh's garden.

Lack of building materials after the war meant that the burgeoning space requirements of the EGI and the BAP could not be immediately met. The temporary solution, worked out by the university registrar Douglas Vale, probably over the objections of the residents of St. Hugh's, was to utilize the recently vacated Nissen huts in the St. Hugh's garden. Therefore, in April 1947, the EGI and the BAP moved to adjacent huts, after renovations that made them more suitable as office, laboratory, and library space. A third hut was occupied by the Laboratory for the Design and Analysis of Experiments, headed by David J. Finney, who had been appointed as the university's first Lecturer in the Design and Analysis of Scientific Experiments in 1945. The huts were connected by concrete paths, which had been constructed so that patients could be wheeled among units.

In July 1947, the university created the Department of Zoological Field Studies, which consisted of the EGI and the BAP. Alister Hardy, Head of the Department of Zoology and Comparative Anatomy, also served as Head of the new department. The effect of the decision was to tie the two institutes more closely to the university, and it marked the beginning of significant funding from the university for the EGI. The directorship of the EGI also became an official university post, helping to secure the future of the institute.

CANARY ISLAND EXPEDITION

In early 1948, David Lack and Mick Southern of the BAP spent 3 weeks studying the avifauna of Tenerife, the largest of the Canary Islands. They arrived at their base of operations, Puerto de la Cruz, on March 15. Their primary objective was to compare habitat utilization and vocalizations of species on Tenerife with those of the same species in Great Britain. In particular, they wished to test the hypothesis put forward by Lack in *Darwin's Finches* that species on small islands, with reduced numbers of closely related relatives, would expand their use of habitats in the absence of competitors (see Chapter 3). A secondary objective was to compare the clutch sizes of species occurring in both the Canary Islands and Great Britain, but they were unable to achieve this objective because of the unusual lateness of the onset of breeding in the Canary Islands in 1948. During their stay, Lack and Southern visited all parts of the island except the delta of the Anaga Peninsula in the northeastern section.

Lack and Southern's findings from the expedition appeared in a jointly authored paper in *Ibis* in 1949. Their observations supported Lack's hypothesis with data on species of tits, warblers, wagtails, and the common raven, all of which occurred in habitats on Tenerife not occupied in Great Britain and Western Europe. They also found that for one species, the chaffinch, the presence on Tenerife of another species in the same genus, the endemic blue chaffinch, resulted in absence of the chaffinch from pine woodlands that it normally occupied in Great Britain. Lack and Southern also reported that neither the songs nor the call notes of the 16 species occurring both on Tenerife and in Great Britain differed detectably.

NIKO TINBERGEN MOVES TO OXFORD

In September 1949, Niko Tinbergen left his post at the University of Leiden to accept a lectureship in animal behavior at Oxford. Aware that Niko had decided to move to a post in an English-speaking country, both David Lack and Ernst Mayr had been trying to convince him to come to their respective institutions. After his trip to

Oxford in 1946 (described earlier), Niko had proceeded on to New York, where he was hosted for 3 months by Mayr. The proximity of Oxford to his native Holland probably played a key role in his decision to choose Oxford.

Housing was required for Niko, his wife, Lies, and their four children on their arrival, and because of currency restrictions they had been able to bring only a few guilders with them. Peter Hartley located a former pub, The Worcester Arms, in Islip, a village 6 miles north of Oxford, and the Tinbergen family moved into it when they arrived. They found more suitable accommodations a few months later on the Banbury Road in Oxford.

Niko founded the Animal Behaviour Research Group (ABRG) in the Zoology Department shortly after his arrival and quickly began to attract students. One of Lack's students in the EGI, Robert Hinde, worked closely with Tinbergen during the final year of his doctoral program. Tinbergen's first group of students, who became known as the "Hard Core," included several who made significant contributions to the field of animal behavior on both sides of the Atlantic. One of these was an American, Martin Moynihan, who coined the "Hard Core" appellation and returned to the United States to become a prominent ornithologist and the founding director of the Smithsonian Tropical Research Institute in Panama. Others included Mike Cullen, a seabird biologist who remained at Oxford for many years as Tinbergen's right-hand man; Desmond Morris, author of *The Naked Ape* (McGraw-Hill, 1967); Aubrey Manning, a zoologist at Edinburgh University and a well-known broadcast naturalist; and Hans Kruuk, a Dutch student who subsequently was cofounder of the Serengeti Research Institute in East Africa and Tinbergen's biographer (*Niko's Nature*, Oxford University Press, 2003).

Although the ABRG was not located next to the BAP and the EGI, there were frequent interactions among members of the three groups. When he first arrived in Oxford, Niko often shared pub lunches with both David Lack and Mick Southern. Niko and a number of his students worked on birds, leading to further interactions between the ABRG and the EGI.

MIGRATION STUDIES

One of David's most persistent interests was the study of bird migration, no doubt triggered in part by his wartime involvement in radar (see Chapter 2). His early research efforts were in visual migration, however; only later did he return to radar studies (see Chapter 8). During late summer, he frequently traveled to the coast to observe arrivals and departures of birds migrating from or to the continent across the English Channel, as well as daily and weather-related movements of the common swift (see Chapter 6). He and his new wife, Elizabeth, spent the early autumns of 1949 and 1950 observing migration through passes in the Pyrenees Mountains of southern France (see Chapter 6). Lack also quantified southward movements along the eastern coast of England.

David's interest during this time in the effects of weather and physical barriers on bird migration extended to the Mediterranean Sea and the Sahara Desert. Reg Moreau was also studying migratory routes of birds from Europe into Africa, and both men wanted to know whether small songbirds crossed the Sahara in a broad front, as was known to be the case for the Mediterranean, or whether their southward movements were confined to eastern and western routes around this formidable barrier. To attempt to answer this question, an expedition into the heart of the Sahara Desert of Libya was organized in the fall of 1953. David Snow, who had just completed his doctorate, and Aubrey Manning, one of Niko Tinbergen's D.Phil. students in the ABRG, participated in the expedition, which was led by Snow. Snow later described their arduous 5-day journey in a Fiat truck from Tripoli to a sprawling oasis near Sebra Fort, almost 1000 miles south of Tripoli. Because they paid the driver £5 each for the trip, including food, they were able to sit in the cab next to the driver while several other passengers rode on top of the merchandise in the open-topped truck. The pair spent 3 weeks at the fort, an outpost of the French Foreign Legion, living comfortably in the civilian mess. They spent their days at the oasis looking for migrants, which proved to be numerous:

> At first, in mid-September, shrikes, Golden Orioles, Hoopoes [were] among the chief species. By the end of our stay, approaching mid-October, these had gone, and Willow Warblers, White Wagtails and Tree Pipits had appeared. Some, such as Swallows and Redstarts, were with us the whole time.[27]

Their observations clearly demonstrated the presence of a broad-front migration of small birds across the Sahara.

DISSOLUTION OF THE EGI/BTO RELATIONSHIP

The incorporation of the EGI into the university initiated changes in the relationship between the EGI and the BTO. These were signaled by David's second annual report of the activities of the EGI, which began as follows:

> Big changes are going forward in regard to administration of the Edward Grey Institute at Oxford. It will be remembered that the foundation of an institute

of field ornithology at Oxford was the primary reason for the creation of the British Trust for Ornithology, as the University was not prepared to accept financial responsibility for such an institute.[28]

Before incorporation of the EGI into the Department of Zoological Field Studies in July 1947, oversight of the institute had rested with the University Committee for Ornithology, on which the BTO had two members. With the creation of the new department containing the EGI, the committee became "anomalous" (as Lack described it), and it was abolished. Administrative responsibility now rested with the head of the department, to whom the director of the institute would report. Significant university funding also meant that the institute did not rely as much on funding from the BTO. Before the change, the BTO had provided about two-thirds of the institute's budget, whereas after the change it provided less than 10 percent of the greatly expanded budget.

The EGI and the BTO continued to share quarters in Oxford until July 1951, when the BTO moved into its own offices. The relationship between the two organizations grew increasingly fractious from 1945 until 1951, a conflict that was perhaps inevitable as the differences in the expectations of the University and those of the BTO played themselves out as David Lack began to place his personal stamp on the EGI.

Lack envisioned the role of the EGI to be that of a national center for the professional, scientific study of birds in the field. This view deviated substantially from the original conception of the EGI as a partner of the BTO in the coordination of national censuses of birds conducted largely by amateur ornithologists. The major protagonists in the conflict were Lack and Max Nicholson, who became chairman of the BTO in 1947. In correspondence with Nicholson, Lack expressed disappointment in the quality and productivity of BTO research efforts and indicated that he did not want the EGI to be associated with such efforts. Nicholson agreed in part with this assessment but continued to uphold the original vision of cooperation between the two organizations as a valid one. In November 1949, Nicholson apparently approached Alister Hardy, chairman of the Department of Zoological Field Studies, with a proposal to remove Lack as EGI Director, citing Lack's failure to adequately support the vision of EGI/BTO collaboration. Hardy rebuffed this suggestion, no doubt reflecting Oxford University's interest in having the EGI be a center of research and scholarship. Lack's transformation of the EGI was thereby validated, and the schism between the EGI and the BTO continued to widen.

In February 1950, Lack submitted a proposal to the BTO Council that included four points: (1) EGI staff should not have special positions on the BTO

Council or committees (in particular, Professor Hardy was prohibiting the service of the Director as an *ex officio* member of BTO committees); (2) the EGI should not expect a grant from the BTO; (3) the EGI needed additional space and would therefore ask the BTO to find new accommodations by April 1950; and (4) BTO members would retain their access to the Alexander Library and could establish a duplicate library at the BTO. The council accepted provisions (1), (2), and (4), and took note of (3). However, the physical separation did not occur until July 1951. The physical separation of the EGI and the BTO mirrored the continuing divergence in their goals and missions and increased when the BTO moved from Oxford to Beech Grove, Tring (Hertfordshire) in 1963. The BTO subsequently moved to its present location in The Nunnnery, Thetford (Norfolk), in 1991.

The BTO established the Tucker Medal in 1954 in honor of Bernard Tucker, one of the founders of the Oxford Bird Census, precursor of the EGI and the BTO. The medal honors individuals who have made outstanding contributions to the Trust's scientific work, through Trust surveys or Trust-aided investigations. Evidence of the degree to which the two offspring of the Oxford Bird Census became separated is the fact that in the almost 60 years since the first medalist was named, only two individuals with direct ties to the EGI have been honored—W. B. Alexander, first Director of the EGI, on the occasion of his "second retirement" in 1955, and John Gibb, one of Lack's first doctoral students, the following year.

MOVE TO THE OXFORD BOTANIC GARDEN

My favorite approach to old Oxford is the coach ride from London or from one of the major airports, Heathrow or Gatwick. The bus travels through Headington, down the hill past Oxford Brookes University into St. Clement's Street and around The Plain into High Street. As you proceed up the High toward the old city and pass over the narrow floodplain of the River Cherwell, your eyes are drawn to the picturesque Magdalen Bridge over the Cherwell and to the dramatic tower of Magdalen College on the right side of the High, just across the bridge. Out of the corner of your eye, you may glimpse the punts huddled against the bank on the west side of the Cherwell. Because of these captivating and awe-inspiring views, you are liable to miss seeing the yellow sandstone buildings on the left side of the High, across from Magdalen—the Oxford Botanic Garden. The buildings, which sit a few feet below street level and are set back about 50 feet, are fronted by a magnificent rose garden. An arched entryway to the gardens is located to the west of the two wings of the building. In January 1952, when the Department of Botany vacated the buildings to move to new quarters on South Parks Road, the west wing became the new home of the EGI and the BAP.

More agreeable accommodations for field-oriented biologists are difficult to imagine. Certainly the Botanic Garden was a far more appropriate home for them than the concrete monstrosity into which the EGI moved in 1971, where it still resides (see Chapter 12). Stone walls completed in 1633 surround the original garden, Britain's oldest botanic garden, which has been extended westward along the banks of the River Cherwell. The original purpose for establishing the garden was as a repository for medicinal plants, and this objective is still reflected in the present-day garden. Jan Morris's description captures something of its mystique:

[W]ith its crumbled stones and shaded benches, its urns and pots and greenhouses, the Cherwell flowing sweetly beside its lawns, and the goldfish who twitch in its ornamental pond—with the great tower of Magdalen serene above its gate, and the spires peering always between its foliage, there can be few better places in England for the contemplation of flowers.[29]

Or, presumably, for the contemplation of birds and other animals.

The EGI and the BAP occupied adjacent banks of offices on the second floor of the west wing of the building next to the main gate, with the EGI inhabiting the offices closer to the gate. Infamously, the door separating the two was always locked, indicative to some of a strained relationship between Lack and Charles Elton, Director of the BAP.

BRECKLAND RESEARCH UNIT

In 1952, the EGI received a 5-year grant of £2000/year from Nature Conservancy to study the relationship between birds and their food sources in state-owned conifer plantations. The study focused primarily on tits and was conducted by two of Lack's original D.Phil. students, John Gibb and Monica Betts, both of whom were appointed as research officers in the EGI after completion of their doctorates.

INTERNATIONAL MEETINGS AND OTHER INTERNATIONAL TRAVEL

Three of Lack's early trips abroad after his appointment as Director of the EGI have already been described: his trip to Holland in February 1946, his participation in the Princeton conference in January 1947, and his trip to Tenerife in 1948. In 1950, he attended the X International Ornithological Congress, held in Uppsala. As its name implies, this meeting was established to bring together ornithological researchers from throughout the world. The first Congress was convened in Vienna in 1884, and it was held at irregular intervals thereafter until 1926, when a 4-year interval was adopted. World War II and its aftermath resulted in a 12-year hiatus, and the Uppsala gathering was its first post-war meeting.

Lack was invited to present one of the plenary addresses, and he chose as his topic the focus of his research at the EGI, population ecology of birds. Many of the themes covered in the address presaged the ideas he later presented in *The Natural Regulation of Animal Numbers*.

BERNARD TUCKER DIES

The midpoint in David Lack's first decade as Director of the EGI was marked by the passing of one of the most influential individuals in the development of academic ornithology at Oxford, Bernard Tucker. Tucker, who had taken First Class honors in zoology at Magdalen College, Oxford, in 1923, had come to Oxford in 1926 as University Demonstrator and Lecturer in Zoology and Comparative Anatomy after 1 year as Demonstrator in the Zoological Laboratory at Cambridge. Professor Stanley Gardner had intended to appoint him to Hans Friedrich Gadow's position as Lecturer in Vertebrate Anatomy there, but Gadow had deferred his retirement, and Tucker returned to his alma mater. While Tucker was an undergraduate at Oxford, he had served as the first honorary secretary of the Oxford Ornithological Society when it was founded in 1921. He was again active in the Society when the Oxford Bird Census was initiated in 1928. He was a strong supporter and close friend of David Lack after Lack's appointment as Director of the EGI in 1945.

At the banquet at New College during the Jubilee Student Conference in Bird Biology in January 1997, the banquet speaker, Robert Hinde, related a bizarre story involving Tucker, Lack, and the pioneering avian physiologist, A. J. "Jock" Marshall. Marshall, a colorful Australian who had lost an arm while fighting behind enemy lines in New Guinea during World War II, was appointed Demonstrator in Physiology at Oxford in 1947. At tea time one day 2 years later, after learning that his appointment had not been renewed, he approached Tucker and Lack, whom he blamed for the decision, and performed a New Guinean dance around the two, complete with costume, thereby placing a curse on them. As Hinde described it, within a year Tucker was dead and Lack was struck with Bell's palsy, a paralysis of one side of the face that is the result of damage to the seventh cranial nerve. Lack never recovered full mobility in the right side of his face, and as a result he had a crooked smile that is evident in many of the photographs taken after 1950 and was interpreted by some as a supercilious sneer. Hinde left his listeners to draw their own conclusions about the incident.

After leaving Oxford in 1949, Marshall spent 11 years at St. Bartholomew's Hospital Medical College, London, before returning to Australia as Professor in Zoology and Comparative Physiology at Monash University. He played a significant role in the development of an outstanding program there and also hired one of Lack's doctoral students, Doug Dorward (see Chapter 10). Jock died in 1967.

FIRST AMERICAN JOINS THE EGI

Although a number of Americans had visited the EGI over the years, including a brief sojourn by Nicholas and Elsie Collias in the summer of 1949, the first American formally appointed to the EGI arrived with a Fulbright fellowship after completing his undergraduate degree in zoology at the University of Wisconsin. David Lack actually held a low opinion of the quality of American education and had been reluctant to accept American students. Gordon Orians, however, came with the strong recommendation of John Emlen, whom David had met and worked with during his sojourn at the California Academy of Sciences in 1939 (see Chapter 3). The acquaintanceship had been renewed when Emlen and Aldo Leopold invited David to speak at the University of Wisconsin after his participation in the Princeton conference in the winter of 1947.

Orians arrived in October 1954 and received an appointment as a research student, a position that he held for 1 year before returning to the United States. His principal research responsibility during his tenure with the EGI was analysis of the population dynamics of Manx shearwaters on Skokholm Island, based on recaptures of ringed shearwaters. In this endeavor, he received considerable assistance from P. H. Leslie, a mathematical ecologist at the BAP. Collaboration between students at one of the institutes and senior individuals at the other was not uncommon, and theirs resulted in a paper authored by Orians with a statistical appendix by Leslie (*Journal of Animal Ecology*, 1958).

Orians recalled two very significant interactions with Lack that occurred during his year at the EGI, one professional and the other personal. At his first tutorial with Lack shortly after his arrival, David asked Orians whether he had written anything that he could read. Orians replied affirmatively, having written a senior thesis at Wisconsin on the food of breeding red-tailed hawks and great horned owls that had been well received by his professors. He was shocked and deeply disappointed at his second tutorial when the thesis was returned covered with red ink, especially when

Lack informed him that he had not corrected any grammatical mistakes or spelling errors but only problems in the logic and cogency of his arguments. David then spent an hour explaining the necessity of thinking evolutionarily and developing testable hypotheses—in short, elucidating the hypothetico-deductive methodology of science. Orians returned to his office after this tutorial crestfallen and certain that he was not destined to be a scientist, but on reflection he realized that Lack's critique was correct, and he resolved to learn from the embarrassing episode. He learned his lesson well. After returning to the United States, he obtained a Ph.D. in zoology at the University of California, Berkeley, under Frank Pitelka and had an outstanding 35-year teaching and research career at the University of Washington. He is widely recognized as one of the pioneers of behavioral and evolutionary ecology in North America and was elected a member of the National Academy of Sciences (USA) in 1989, a member of the American Academy of Arts and Sciences in 1990, and Eminent Ecologist by the Ecological Society of America in 1998. Orians still marks that unhappy tutorial with Lack as the beginning of his transformation into becoming a scientist.

Orians' personal experience with Lack was equally long-lasting but considerably more pleasant. On June 25, 1955, Orians married Elizabeth Newton in the chapel of Lincoln College, Oxford. They had met while both were students at the University of Wisconsin, and she had sailed across the Atlantic after completing her studies so that they could honeymoon in Scandinavia after their wedding. Neither set of parents could make the trip for the ceremony, so David Lack acted as a stand-in for Elizabeth's father, walking her down the aisle of the chapel. After their return from the Scandinavian honeymoon, David took them on a day trip to the Cotswolds, the "Heart of England." The union lasted for more than 55 years.

W. B. ALEXANDER RETIRES AGAIN

In 1955, at the age of 70, W. B. Alexander, the first Director of the EGI, retired from his position as volunteer Librarian at the institute. Alexander had founded the library in 1930 with a small collection owned by the Oxford Ornithological Society, even before the inauguration of the institute. In 1947, the last act of the Committee for Ornithology was to name the library for him—the Alexander Library of Ornithology, a name that it retains to this day. At the time, Alister Hardy described it as "one of the most extensive ornithological libraries in the world for works published after 1900."[30] During the 10 years he served as volunteer Librarian, Alexander worked tirelessly and successfully to enhance the library's collection, not only adding many ornithological journals and books but also amassing a prodigious collection of reprints. This reprint collection, arranged taxonomically and topically, remains a valuable means of access to older literature that is otherwise difficult to acquire. At the time of Alexander's death in 1965, the author of his obituary described him as "knowledgeable yet not pedantic, critical yet sympathetic to the most callow of opinions, . . . the ideal teacher."[31] David Lack heralded the Alexander Library as probably the most complete ornithological library in the world and credited Alexander with its development.

Alexander's two retirements represented fitting ellipses to the first 10 years of David Lack's tenure as Director of the EGI. During this period, he had placed his stamp on the EGI, establishing it as an international center for the study of birds. His first crop of doctoral students, including Robert Hinde, Monica Betts, David Snow, John Gibb, James Lockie, and Bill Sladen, had all completed their degrees by 1955 (see Chapter 11). The book for which this chapter is named had been published. The next decade of Lack's directorship would see a new crop of students and changes in the direction of his research focus, as well as consolidation and expansion of the programs he had already inaugurated (see Chapter 8).

The mutual preening of swifts . . . perhaps helps to inhibit potential hostility between the pair, or to cement the pair-bond, and so may be compared with the affectionate behaviour of man and wife. . . . [T]he mutual preening become[s] much less excited in the latter part of the breeding season, in the same way that the delighted greeting of the newly-wed gives place to the casual 'Hello' of later and more contented years.

—*Swifts in a Tower*, pp 42–43

6

Swifts in a Tower

SWIFTS IN A *Tower*, published by Methuen in 1956, was based on 10 years of field work on the common swift at Oxford. The first 2 years of observations were performed in two villages near Oxford, but in 1948 the study was shifted to the tower of the University Museum of Natural History. Nest boxes with glass backs were placed inside the vent openings in the tower to facilitate direct observation of the swifts. Most of the original observations on the swifts by Lack and other personnel from the Edward Grey Institute of Field Ornithology (EGI) were made in the tower of the museum—hence the name of the book. *Swifts in a Tower* not only covered the biology of swifts with particular emphasis on the common swift, but it was also a treatise on Darwinian natural selection, focusing on the many adaptations of swifts for life in the air.

The University Museum is located on Park Street opposite Keble College near the southwest corner of The University Parks. It was built of variegated brick and stone between 1855 and 1860 in a neo-Gothic style. Lack described the laying of the cornerstone for the museum:

The story begins on June 20th, 1855, when a group of bearded and reverend scientists, a combination now, alas, unknown, joined to sing the Benedicte in the open air at the edge of Oxford, as the foundation stone was laid for a re-markable new building, Oxford's University Museum of Science. And as they sang 'O, all ye fowls of the air, bless ye the Lord' there came, it may be supposed, an answering scream from the circling swifts, for in the top of the new building, these birds would find their home in the years to come. . . . The spirit in which the new building went forward is shown by the prayer composed for the foundation ceremony by the Professor of Medicine, Sir Richard Acland. 'Grant that the building now to be erected on this spot may foster the progress of those sciences which reveal to us the wonders of Thy creative powers. And do Thou, by Thy heavenly grace, cause the knowledge thus imparted to fill us with the apprehension of Thy greatness, Thy wisdom, and Thy love.' In those days scientists knew that God was glorified in His works. There was no forewarning shadow of the great debate on Darwinism, to be held in the scarcely completed building only five years later.[1]

Lack would recount the particulars of that debate in his next book (see Chapter 7), but the focus of *Swifts in a Tower* was a description of the biology of the group of birds that he claimed were the most adapted for life in the air of any birds. Like *The Life of the Robin*, the book was written for the general reader with numerous literary and historical references included and a paucity of technical jargon. As with the earlier book, however, it was based on a rigorous scientific understanding of the life of swifts. A new edition of *Swifts in a Tower* was published in hardback in 1973 by Chapman and Hall, which also published a paperback edition in 1979.

Several early chapters dealt with detailed descriptions of the breeding biology of the common swift, including acquisition and defense of a nest site, courtship, egg laying and incubation, nestling growth and fledging, and age at first breeding. Common swifts typically breed for the first time as 2-year-olds, which means that there are many nonbreeding birds in the breeding colonies. A chapter on swift nests included a description of the remarkable nest sites and nests of different swifts from throughout the world, including the saliva nests of several species of cave swiftlets in Southeast Asia:

> The nest is nearly white and looks rather like water-glass, and it is one of the few birds' nests of commercial value, as it is the main ingredient of birds' nest soup, prized by the Chinese.[2]

The common swift is almost exclusively an aerial insectivore, capturing insects on the wing to feed the developing young. Lack described another interesting adaptation of the species that enables the young to survive for extended periods when flying insects are scarce due to cold or rainy weather. In such conditions, the young can enter a state of torpor and suspend development for several days, resuming growth and development when food becomes available again. An interesting omission from the book was its failure to cover one of Lack's most original theoretical contributions, the brood reduction hypothesis related to hatching asynchrony (see Chapter 5).

Lack developed this hypothesis to account for the early onset of incubation in the common swift and the resultant asynchronous hatching in the species.

Later chapters described some of the remarkable adaptations of the common swift to life in the air. These include their habit of roosting on the wing. At dusk the nonbreeding swifts spiral upward until they are out of sight, and in coastal areas they actually fly out to sea to spend the night in the air. A second adaptation is the avoidance of inclement weather and poor feeding conditions by undergoing weather-related movements. Large numbers of swifts, again mostly nonbreeding individuals, fly long distances away from advancing fronts, returning after the cold, wet weather has passed. These flights may include movements across the North Sea or the English Channel. Another remarkable adaptation of swifts is their ability to copulate on the wing, an observation first reported by the renowned English naturalist Gilbert White in 1795. Swifts usually copulate at the nest site, and at the time of Lack's work on swifts, descriptions of aerial copulation were uncommon. Dennis Chitty recalled meeting an ebullient David Lack one day outside the museum: Lack excitedly described his experience of witnessing aerial copulation, an incident whose significance was emphasized by its contrast with Lack's normally reserved demeanor.

In the final chapter of *Swifts in a Tower*, "The Meaning of Adaptation," Lack discussed the implications of the Darwinian revolution for those who prayed, with Sir Richard Acland, "Grant that the building now to be erected on this spot may foster the progress of those sciences which reveal to us the wonders of Thy creative powers." While the common swift with its remarkable adaptations to life on the wing provides some extraordinary examples of adaptation, Lack pointed out that all organisms are adapted to their particular way of life: "After marveling at swifts and eagles, gazelles and elephants, let us remember that the world is inhabited chiefly by sparrows and rats."[3] He asserted that evolution by natural selection successfully accounts for these adaptations, which previously had been cited by William Paley and others as an argument for design and, consequently, postulation of a Designer. Lack concluded, "After a bitter struggle, the occurrence of evolution came to be accepted as proved, even for man,"[4] but he then proceeded to discuss the difficulty of rationalizing our understanding of man's nature with a wholly naturalistic explanation, a topic that would be central to his next book (see Chapter 7). He concluded *Swifts in a Tower* with a somewhat enigmatic line:

> This is a grave matter with which to end a bird-book, but the tower where our swifts live was built by those who held that the study of nature should lead us, through a truer understanding, to a fuller worship of the Creator, and the times urgently require us to search out the basis of our lives.[5]

Although *Swifts in a Tower* is out of print, a booklet by Andrew Lack and Roy Overall entitled *The Museum Swifts* is sold at the Museum gift shop.

MARRIAGE

Elizabeth Silva was born in Hertfordshire in June 1916, but moved at the age of 3 years to Kent. Her father, Jack, owned a starch manufacturing business in London. She studied violin and piano as a child but was also intensely interested in birds and the out-of-doors from an early age, an interest that she pursued on her own. She was planning to enter the Royal Academy of Music, London, to study piano, but the destruction of her father's starch factory in the early days of the London blitz derailed that plan. Instead, she enlisted in the Auxiliary Territorial Service, serving along with 217,000 other women, including the young Princess Elizabeth. Her principal duties were servicing and driving ambulances, and she served for more than 5 years in England and France until the war in Europe ended. She arrived in Paris shortly after VE Day in May 1945.

Like many other women at the end of the war, Elizabeth found herself looking for work. The two principal occupations open to women at the time were nursing and secretarial work, and despite having limited training, she decided to seek a secretarial position, preferably one that would relate to her interest in birds. She contacted one of the few natural history people she knew of, Richard Fitter, who passed her résumé on to the newly appointed director of the EGI along with a note saying, "Here's another for your reject file." The outcome proved Fitter very wrong indeed! David Lack invited Elizabeth to interview for the position.

Elizabeth arrived back in England on Christmas Eve, 1945, to interview for the job as Lack's secretary, and she obviously impressed the new director. She began work in early January 1946. A few weeks later, as Elizabeth described it, Lack arrived early in the morning at his research site in Wytham Woods and encountered Elizabeth emerging from the wood by herself after an early-morning bird-watching excursion. He promptly invited her to be his part-time field assistant in addition to her secretarial duties, an offer she happily accepted. Her first field duties at Wytham Woods involved the robin, and she proudly recalled that she found a record number of robin nests during the 1946 breeding season. However, David quickly realized that much larger sample sizes were required to answer the questions he was posing, and so Elizabeth's field duties shifted to helping John Gibb and Peter Hartley monitor the nest boxes in the tit studies that began in 1947 (see Chapter 5). Transportation from the EGI to Wytham Woods was usually by bicycle (although Hartley occasionally drove his old red sports car, *Bloody Mary*), and the quantity of field work required an early start and meant that it was often difficult to get back to Oxford before midafternoon.

Another project on which Elizabeth worked was the study of swifts breeding in the tower of the University Museum a few blocks from the EGI. At that time, the EGI was located in the garden of St. Hugh's College on the Banbury Road (see Chapter 5). It is unclear when David's feelings for his field assistant began to change, but it may have been at the time of the following incident that he began to recognize it. On days when she was observing the swifts, Elizabeth would return to her office in the EGI by 1 p.m., when David and other members of the EGI staff met for lunch. One day, David suddenly realized that she had not returned from the tower as usual, and he rushed the half-mile from St. Hugh's to the museum, half expecting to find her crumpled body lying at the bottom of the rickety ladder that led up into the museum tower. Happily, no such accident had occurred. Instead, Elizabeth was observing for the first time a fight between two male swifts for possession of one of the nest boxes. The two birds were locked together, each gripping the other's feet, remaining essentially motionless for hours except for occasionally pecking at each other. More than 60 years later, Elizabeth still spoke with excitement about the event, saying, "I couldn't leave until I found out which male won. This was the first time that anyone had ever witnessed such a fight."[6] Whether or not this was a moment of dawning awareness of a changing relationship, David and Elizabeth became engaged later that year.

Their wedding was a small, family affair, held on July 9, 1949, in Elizabeth's parents' parish church in Underriver near Sevenoaks, Kent. The village, whose name is derived from an old English term meaning "under the hill," is today an affluent residential community surrounding St. Margaret's Church. The day was warm and sunny as the Reverend Jim Wilson, father-in-law of Elizabeth's older sister Mary, officiated. Wilson was the brother of Dr. Edward Adrian Wilson, a member of Scott's ill-fated Antarctic expedition. David's best man was his Cambridge friend and wartime colleague, George Varley, who was by then Hope Professor of Entomology at Oxford. The wedding party was small and consisted primarily of family members and a few close friends. Among the latter were Peter Hartley, a friend of both Elizabeth and David from the EGI, and Mary Neylan and her teen-age daughter Sarah, from Dartington Hall School. Family members attending included Elizabeth's parents, her brother Ted, and two second cousins. Members of David's family included his mother, brother Christofer, and sister Katreen.

Elizabeth recalled that as they drove from Underriver toward their first night's destination in Lavenham, Suffolk, yellowhammers were singing all along the roadside—a fitting tribute to the newlyweds that only two ornithologists would recognize and appreciate.

FIGURE 11: David Lack and Elizabeth Silva on their wedding day, July 9, 1949. (Photograph courtesy of Lack family.)

When shall I see the white-thorn leaves agen,
And yellowhammers collecting the dry bents
By the dyke side, on stilly moor or fen,
Feathered with love and nature's good intents?
Rude is the tent this architect invents,
Rural the place, with cart ruts by dyke side.
Dead grass, horse-hair and downy-headed bents
Tied to dead thistles—she doth well provide,
Close to a hill of ants where cowslips bloom
And shed oer meadows far their sweet perfume.
In early spring, when winds blow chilly cold,
The yellowhammer, trailing grass, will come
To fix and place and choose an early home,
With yellow breast and head of solid gold.

The Yellowhammer (John Clare, 1793–1864)

After spending the night at the Swan Inn in Lavenham, the couple proceeded to the Burnham Overy Staithe windmill on the Norfolk coast adjacent to Scolt

Head. They spent a week there, often walking out onto Scolt Head, which is an island only at high tide, before returning to Oxford. David had first visited the island on a Natural History Society–sponsored field trip during his first summer at Gresham's School, and Elizabeth had also been to Scolt Head camping with members of her family. Nicholas and Elsie Collias, an American couple who were spending a few weeks at the EGI during the summer of 1949, regularly weighed common swift nestlings in the Museum Tower during the Lacks' absence. They, too, were newlyweds, having been married in December 1948. They went on to distinguished ornithological careers in the United States, spending most of their professional lives at the University of California, Los Angeles.

David and Elizabeth's romance was not the only one to flourish in the relatively confined quarters of the EGI and the Bureau of Animal Population (BAP). Peter Crowcroft, who arrived in 1949 as one of Charles Elton's D.Phil. students in the BAP, was working on shrews in Wytham Woods. He met another BAP D.Phil. student, Gillian Godfrey, who was working on voles at Wytham, and they eventually married. Crowcroft, who would later chronicle the history of the BAP in *Elton's Ecologists* (University of Chicago Press, 1991), had come to Oxford from his native Tasmania, where he had been a tap-dance champion known as "Pete with the Dizzy Feet," and had earlier married and divorced a former Miss Tasmania.

A short time later, another romance blossomed, between Bill Sladen, one of Lack's D.Phil. students, and Brenda MacPherson, field assistant to Elton, and they also married. Another of Lack's D.Phil. students, N. Philip Ashmole, who arrived in 1957, fell in love with the new EGI librarian, Myrtle J. Goodacre. Ashmole spent much of the next 2 years on Ascension Island as part of the Oxford expedition there, but the couple married in 1960 after his return. They later spent half a century together studying the ecology of the Ascension and St. Helena Islands.

Yet another romance developed, between a long-time BAP researcher, Mick Southern (see Chapter 11), and Kitty Paviour-Smith, a research student who had come to the BAP from the University of Otago in New Zealand. Their relationship literally blossomed in Wytham Woods, where both were conducting research. The first indication of their romantic involvement came when another researcher entering Wytham surprised them walking hand-in-hand leaving the Woods. Mick was already married, but he soon divorced his wife and married Kitty. Finally, although it was far from the last romance to flourish within the EGI in the years since, Chris Perrins, one of Lack's doctoral students who later succeeded him as Director of the EGI, courted another EGI librarian, Mary Carslake, whom he married in 1963.

The Lacks' first home was a first-floor flat at 40A Park Town (not 100 yards from the semidetached house shared by Charles Elton and his wife, Joy Scovell).

After Niko Tinbergen and his wife, Lies, arrived in Oxford in late 1949, they settled in a house on the Banbury Road a few blocks from the Lacks' Park Town flat. David and Elizabeth often entertained the Tinbergens, both of whom were good friends. The same was not true for Elton and his wife, however. Charles and David were never friends (see Chapter 11), and Joy was, in Elizabeth's words, "painfully shy."

A few months after their first son, Peter, was born, David and Elizabeth moved to Heath Cottage, one of a row of four attached cottages along The Ridgeway on Boars Hill. Boars Hill, which lies 3 miles to the southwest of central Oxford, is sometimes referred to as the home of poets, although the poet most frequently associated with the hill actually never lived there. Heath Cottage is only half a mile from the vantage point on Boars Hill that inspired Matthew Arnold's description of Oxford in his poem, *Thyrsis*, as "that sweet city with her dreaming spires." Two successive Poets Laureate (Robert Bridges and John Masefield), Robert Graves, and other poets did reside there at different times. Although known for its poets, Boars Hill had also been the home of at least one other pioneering scientist, Sir Arthur Evans. An archeologist, Evans is best known for his excavation of Knossos on Crete and for his development of an understanding of Minoan civilization.

When the Lacks purchased Heath Cottage in 1952, the property included a 1-acre tract behind the four cottages on which they intended to build a home. Construction of the house, known as Hatherton, began on the day their second son, Andrew, was born in late 1953, and David and Elizabeth moved in, along with their two sons, in August of 1954. Hatherton was somewhat smaller than they had hoped because postwar restrictions were still in effect, limiting the size of new homes to a certain number of square feet per person.

A narrow gravel lane just beyond the four cottages leads from The Ridgeway to a gate surrounded by a bank of thick Lawson's cypress concealing the house from view. On entering the gate, one sees a charming two-story, stucco-sided house with a tiled roof. A large lawn stretches to the south of the house. The lawn is surrounded by trees, a few planted by David, which are now at least 40 feet tall. It is here that the Lacks lived together and raised their four children until David's death. In 2013, Elizabeth was still residing there, nearly 60 years after its construction.

THE FRENCH PYRENEES

Two months after their wedding, David and Elizabeth took their first holiday vacation together in southern France, where Elizabeth's Aunt Fan Twemlow had lived for many years. They spent much of their time observing birds, establishing a pattern that would continue during subsequent holidays.

They returned to the Pyrenees to study migration the following year, a vacation that led to observations that David described as follows:

The most remarkable days for a naturalist combine grandeur with novelty, the beautiful with the rare or unexpected. As a boy, such experiences came to me seeing for the first time a new kind of bird. . . . As I grew older, such memorable days became much rarer, for though the beauty was still there, the unexpected was gone. . . . But there was one much later occasion, just after my fortieth birthday, when in lovely autumn weather amid superb scenery, a deeply impressive spectacle was combined not merely with the knowledge that no one had written of it before, but that one of the puzzles of migration was solved. This happened on October 13th, 1950.[7]

David and Elizabeth had left the French village of Gavarnie at dawn on that date for a 4-hour hike up to the Port of Gavarnie, a narrow pass (less than 50 m wide) at an elevation of just under 2300 m. Before their arrival at the Port of Gavarnie that day, it was widely believed that small songbirds avoided mountains during migration, flying around rather than over them; but what David and Elizabeth observed shattered that belief. They saw hundreds of birds belonging to several songbird species flying southward through the narrow pass only a meter or two above the ground, against a prevailing wind. Even more startlingly, they observed thousands of insects from three orders (Diptera, Odonata, and Lepidoptera). The insects included several species of butterfly, a hawkmoth, one species of dragonfly, and a species of hoverfly, all flying west-southwest through the pass, like the birds against the wind, but flying just a few centimeters above the ground. Migratory movements had been observed previously in some of the species, but this flight through a high mountain pass was a novel observation. The couple published their findings in a jointly authored paper (*Journal of Animal Ecology*, 1951).

CHILDREN

David and Elizabeth had four children: Peter (born January 17, 1952), Andrew (November 2, 1953), Paul (August 29, 1957), and Catherine (July 15, 1959). Elizabeth left her position as a field assistant at the EGI shortly before Peter's birth, and she became the primary caregiver for the children. The family developed few social contacts outside the home. As one of the children recalled:

We were an insular family, keeping ourselves to ourselves. Both my parents are naturally shy people. Relating is an effort. On holidays if there were people in one direction we went in the other. On days out the same.[8]

David was committed to being a more engaged parent than his own parents had been (see Chapter 1), a commitment that he at least partially fulfilled. Elizabeth described him as "conscientious" about his family, a view that was shared by some of David's later D.Phil. students, who said that he was interested primarily in two things, his work and his family. His children remembered, however, that the two commitments were often in conflict. They recalled the staccato sounds of his rapid, two-finger typing emanating from his home study on many weekend afternoons, and Elizabeth sometimes admonishing them to be quiet, "Daddy is working." In fine weather, David would sometimes even do his typing in the garden.

David was devoted to his children, however. The children remembered him as a warm person who spent time with them, talking, reading, and playing games. George Lambrick, who lived next door, recalled that David taught Peter, Andrew, and him to play "golf croquet," a collaborative as opposed to competitive form of croquet that David had learned at Trinity College after his appointment there (see Chapter 8). Lambrick, now an archeologist who lives in what was his parents' home next door to

FIGURE 12: Photograph of Lack family. Seated from left: Paul, David, Catherine, Elizabeth. Standing from left: Peter, Andrew. (Photograph courtesy of Lack family.)

Hatherton, also recalled that he and the Lack brothers played cricket on the lawn of the Lack home, once breaking a window with an errant batted ball. No serious repercussions ensued, even though there were sandwiches prepared for a party that afternoon on a table inside the house that had to be remade after the unhappy accident.

The younger Lack children also had playmates living nearby. The children of David Nichols, an echinoderm biologist who lived from 1957 to 1965 in Gorselands, the fourth of the four cottages along The Ridgeway and the closest to Hatherton, were about the same ages as Paul and Catherine, with whom they often played. The two older Nichols girls once received a "bird-imitator, which you fill with water, then blow through them and they warble."[9] When they played it out their window one day, both David and Elizabeth emerged from Hatherton with binoculars to see what rare bird was visiting their garden. (David kept a list of birds spotted in the garden, and George Lambrick recalled the excitement at Hatherton in 1961 when a red kite was sighted flying over the garden.[10]) On this occasion, however, no exotic bird was seen, and the girls discretely disappeared from their window. Nichols, who took a professorship at the University of Exeter in 1969, remembered his neighbor as friendly but also protective of his privacy.

Privacy, perhaps even isolation, seemed to characterize the Lack family. David and Elizabeth rarely entertained guests. When the children were small, they occasionally hosted visiting ornithologists. A neighbor recalled visits by Roger Tory Peterson, James Fisher, Hugh Cott, and Bernard Stonehouse (who showed his film of Adelie penguins after a meal). The Lacks did less entertaining as the children grew older, however, and the children do not remember any guests for either lunch or supper, nor for Sunday dinner, which was an important family occasion. Although the Tinbergens and Lacks had gotten together frequently when David and Elizabeth lived in Park Town and the Tinbergens nearby on Banbury Road, the children do not remember their being entertained on Boars Hill. Meals were thus strictly a family affair, and David sometimes introduced a topic for family discussion. Foreign colleagues who were also good friends occasionally stayed at the Lack home when they visited Oxford. Ernst Mayr stayed at least once, as did Robert and Betsy MacArthur during a brief visit to Oxford in 1969.

At Christmas, David's sister Katreen usually came from London, where she was a teacher, to spend 2 or 3 days. David took the children carol singing at houses all along The Ridgeway each year, and they collected donations for the Church of England Children's Society, with little Catherine holding the tin.

Evenings in the Lack home, which had no television set, were sometimes spent playing card games, particularly Old Maid. When the children were young, David would often read to them. He required almost 9 hours of sleep a night, a fact he often decried, so evenings usually ended at 9:15 p.m., when he went to bed. It was

David, however, who would respond if one of the children called out in the night. Because Elizabeth cared for the children all day, he believed that he should relieve her of nighttime parental duties. As one of the children said, "though I have to admit that I would have preferred mother to come—she would have been more sympathetic."[11]

Family holidays, like David and Elizabeth's first two visits to the Pyrenees, were spent at places where David could pursue some research interest and watch birds. Typically the family would have two holidays a year, one lasting 10 to 14 days at Easter and a second lasting about 2 weeks in late summer. They would stay in a small guest house with full board, thereby ensuring that Elizabeth had a break from her cooking duties. Destinations over the years included Suffolk, Norfolk, Wales, East Anglia, Dorset, and even Scotland. Despite the fact that they sometimes felt that

FIGURE 13: Lack family on Portland Bill during a holiday in 1967. Portland Bill is an island with a bird observatory off the Dorset coast. (Photograph taken by Paul Lack, courtesy of Lack family.)

they were competing with birds for their father's attention, the children remembered their holidays as "brilliant." Although David typically devoted his mornings to making observations, the family usually spent afternoons at the beach. One of the children commented, "It was not until I went away to school that I learned that not every family went somewhere on holiday to watch beastly birds."[12] The children's first trip abroad was another such excursion. In April 1965, the family traveled to the Camargue, the delta of the Rhone River in southern France and an area well-known for its abundant bird life. Most of the family's spring holidays in the late 1960s were spent in the Scottish mountains or somewhere in Europe, with bird watching being a major focus.

Another frequent "vacation" destination was the island of Skokholm, where some of David's D.Phil. students did their research. David would go alone or with one or both of the older boys. One of his students recalled that when David arrived on the island, he said that the student should not expect to see much of him, and he didn't.

All of the children attended Dragon School in North Oxford as day students. David would sometimes drive them to school on his way to work or pick them up on his way home. The Dragon School campus borders the River Cherwell, scarcely 100 yards from the Lacks' first flat in Park Town. It was founded by Oxford dons in 1877 as a coeducational preparatory school with both day and boarding students, the latter category restricted to boys until 1994.

The boys all entered Bryanston after finishing preparatory school. The choice of Bryanston, a progressive public school located in Blandford Forum, Dorset, reflected David's belief in progressive education, a belief that had been shaped by his experiences at Gresham's School (see Chapter 1) and Dartington Hall School (see Chapter 2). Catherine attended Bryanston, with its emphasis on Pioneering, only for the sixth form, after first attending Oxford High School for Girls. She was something of a pioneer herself, being among the first girls to attend the school when it made the transition to coeducation in 1972. It was not the last time she was to be a trail-blazer.

After finishing his studies at Bryanston in December 1969, Peter spent 7 months in Tanzania and then most of the next academic year in Jamaica, while David was on sabbatical there (see Chapter 12). He had developed an interest in birds early in childhood and actually assisted his father and Tony Diamond, David's field assistant in Jamaica, with their field work on the birds of the West Indies. He read zoology at St. John's College, Oxford, completing his B.A. in 1974, and then joined the EGI as a doctoral student in ornithology. He spent more than 2 years collecting data for his thesis at the Tsavo East National Park in Kenya, and when he completed his D.Phil. in 1979 under Chris Perris' supervision, he became not only David Lack's biological son but also his "academic grandson." Following completion of his doctorate, he worked for a short time for the Game Conservancy and as summer warden at the Isle of May Bird Observatory before joining the British Trust for Ornithology (BTO) in September 1980. He worked for several years as an organizer of the *Atlas of Wintering Birds* and on a research project on farmland birds before taking charge of the BTO's computer services for 20 years. In 2012, he held the position of Information Services Manager of the BTO in Thetford.

Andrew remained at Bryanston for his final year (A levels) and therefore spent only holidays with the family in Jamaica. He then attended Aberdeen University in Scotland, completing his B.Sci. in botany in 1976. He next proceeded to King's College, Cambridge, where he completed a Ph.D. in botany in 1980. He obtained a 3-year appointment as a research demonstrator in botany at what is now the University of Swansea in Wales and then received a Natural Environment Research Council (NERC) grant for a further 3 years as a postdoctoral fellow at Swansea. In January 1987, he accepted an appointment at Oxford Polytechnic (now Oxford Brookes University), where he became Senior Lecturer in Environmental Biology.

After completing his studies at Bryanston, Paul followed his brother Peter to St. John's College, Oxford, where he read agricultural science and forestry, completing a B.A. degree in 1979. He then completed an M.Sci. in forestry at St. John's before enrolling in Lincoln Theological College in Lincoln, Lincolnshire. After finishing his theological training, he was ordained in Worcester Cathedral by the Church of England in 1986. He served as curate of 10 villages east of Worcester and then spent 11 years as rector of six small parishes near Tenbury Wells, which he described as "a small town in the middle of nowhere much, where Herefordshire, Worcestershire and Shropshire meet." He retired from the priesthood after being diagnosed with myalgic encephalomyelitis (chronic fatigue syndrome), and in 2012 he was teaching adult numeracy (quantitative literacy) and photography part-time in community centers in Worcester and Bromyard.

Catherine read music at Clare College, Cambridge, after leaving Bryanston, and she completed her degree there in 1981. She worked at the Bodleian Library in Oxford and then studied theology at Queen's College, Birmingham, later obtaining an M.A. in Applied Theology at Westminster College in Oxford. She completed her theological training in 1992 and was ordained at Bury St. Edmunds, Suffolk, that same year. She was priested in 1994, being among the first women ordained by the Church of England, again a pioneer. After her ordination, Catherine worked briefly in Suffolk, then as chaplain at Keele University, just west of Newcastle-under-Lyme for several years. She worked a short time at the Yarl's Wood Immigration Centre and as warden of a small retreat house, and in 2009 she was appointed Chaplain of Newcastle upon Tyne University. This post she still held in 2012, also serving as Master of St. Thomas's Church in Newcastle.

Peter, Andrew, and Paul all married, and in 2012 each had one or more children, six among them. David and Elizabeth therefore have six grandchildren, none of whom David lived to see, and one great-grandchild. Catherine has never married.

ILLNESS

Two devastating illnesses struck the Lack family in the span of a few months. Late in 1964, Elizabeth was diagnosed with cancer of the lower right jaw, and early the following year, 5-year-old Catherine developed life-threatening complications after a bout of chicken pox.

Elizabeth spent 2 weeks in hospital for radiation therapy that proved unsuccessful. During her hospitalization, her niece Susan came to look after the house and children. After it was determined that the radiation therapy had been ineffective, and after the trip to the Camargue in April 1965, she entered the hospital again, this time to have her lower right jaw surgically removed. Surgery was followed by more radiation therapy, and the children were cared for by a hired live-in woman for the 3 weeks Elizabeth was hospitalized. This treatment was successful, although the period of convalescence was long. Elizabeth lived for the next 20 years without part of her lower right jaw. In 1987, she was diagnosed with cancer of the palate. After surgical removal of the palate, a replacement plate was inserted, and bone from her hip was used for partial replacement of the lower right jaw. As one of her sons said, "She is quite tough!"

After her case of chicken pox, Catherine developed encephalitis. She was confined to bed for 7 weeks, and every time she lifted her head during that time, she became nauseated. By the time she was able to leave her bed, she had lost the ability to walk. Andrew, who was 11 years old at the time, said, "I remember the rehabilitation well with little walks around the drawing-room floor with my mother."[13] Fortunately, Catherine recovered fully, despite some rather bleak prognoses from her attending physician.

DAVID'S DEBT TO ELIZABETH

Along with his early commitment to being a "kindly schoolmaster" (see Chapter 1), David also wished to be a more engaged parent than his own absentee parents had been, if he ever had children. It is clear that he fulfilled the first commitment during his years of teaching at Dartington Hall School (see Chapter 2). His marriage to Elizabeth and the four children she bore him provided him the opportunity to fulfill the second. How did he do? Judged by the standards of his own parents, he certainly had a greater physical presence in his children's lives, but their own recollections suggest that he may have been emotionally distant from them, despite Elizabeth's strong assertion that he was "conscientious" about his family. One of his children remarked that he "did not 'do' emotional closeness!"[14]

Elizabeth, an accomplished ornithologist in her own right, worked closely with David during his early years at the EGI, particularly on the swift project (described earlier). After his death, she was instrumental in seeing *Island Biology* through to publication (see Chapter 13), and she edited, with Bruce Campbell, the highly regarded *A Dictionary of Birds* (T. & A.D. Poyser, 1985), which was re-released in 2010. Two months before Peter was born, she left her position as field assistant in the EGI to stay at home. She remained a "stay-at-home mom" for the duration of their children's childhoods, providing the principal framework for their upbringing. Elizabeth believed passionately in the significance of David's work and in giving him the freedom to pursue it—which he did. Her exact words were that David was "conscientious about his work, and conscientious about his family."[15] The order of priorities suggested in that statement was real. David was always with his family on holidays, but until the mid-1960s the holidays were in places where he could pursue some aspect of his work. He also frequently worked at home.

As the older children reached their teen years, they began to develop their own interests, Peter in birds and Andrew in plants. David, of course, shared Peter's interest in birds, but Andrew recalled, "He had never been interested in plants at all before, except for the interaction with birds, but took the alpines in particular to heart, and he loved climbing Scottish mountains in their pursuit, mainly because I did."[16] Thus Peter and Andrew were able to share memorable experiences with their father in their teen years. David was less able to relate to Paul's mechanical interests, and Paul and Catherine were too young when he died for them to have the opportunity of sharing their emergence from childhood with their father. Elizabeth had to fill that gap as well, and she did so with great success.

Hence David's enormous debt to her.

OTHER INTERESTS

David's commitment to work and family resulted in a reduced ability to exercise some of his other interests. He had a great love of music, particularly choral music, and while he was at Dartington Hall, he had displayed a strong interest in sports. He enjoyed playing tennis and regularly participated in the annual doubles tournament at Dartington Hall, where a former student remembered him as a fierce competitor in field hockey, playing on the wing. His favorite spectator sport was cricket, and in the early days at the EGI he often took a break after lunch to watch cricket in the University Parks. These pursuits diminished significantly as the number and ages of his children increased, but one of his sons remembered going to the University Parks to watch cricket with his father, who declared: "The Parks is the one place on earth where it is still possible to watch First Class Cricket for free."[17]

David's interest in music never waned, but the opportunities to attend concerts or sung Evensong decreased dramatically. In addition to his love of choral music, he also enjoyed the works of the Classical and early Romantic composers, particularly Mozart, Bach, Handel, Beethoven, and Wagner operas (but only in the opera house). As one of his sons stated,

> Mozart was first and last love—I remember him deciding on almost entirely Mozart operas for his potential 'Desert Island Discs' [for a long-running BBC Radio 4 program that asked distinguished guests to select eight records for a desert island sojourn].... As always decided in his tastes, he did not like grand Italian opera or Mendelssohn (too sentimental!), but had time for Stravinsky, especially neo-classical like *Oedipus Rex*, and Britten.[18]

Andrew recalled that his father occasionally took Peter and himself to evening musical events, including several Mozart operas: "Then he took the best centre-circle seats. Go rarely but go well when you do!"[19]

David also enjoyed art, his favorite era being the post-Renaissance period through the impressionists, but he had no interest in modern painters. He visited the Rijks-museum in Amsterdam with Peter and Andrew, one of whom recalled "my father being very keen on the Vermeers (who isn't?!) and the Rembrandts."[20] When he was in London attending a meeting such as those of the Royal Society, David's first choice for spending a few free moments was a visit to the National Gallery, and the children remembered being taken there on more than one occasion.

Three things that David was decidedly not interested in were politics, housework, and anything mechanical. He had no interest in national politics, and this carried over to an abhorrence of committee meetings, which he considered unproductive and a waste of time. Thankfully, household duties were more than adequately ful-filled by Elizabeth, because David, according to one of the children, could not even wash a saucepan: "He was completely hopeless practicalwise in the house."[21]

Things mechanical also mystified him. He apparently never learned to do electri-cal soldering when he worked in radar during World War II (see Chapter 2), getting a corporal to do it while he went looking for birds. He also found his car somewhat of a challenge. Elizabeth, with her wartime experience in the Auxiliary Territorial

Service, would sometimes have to go out and start the car when David couldn't get it started in the morning.

David's car may deserve a note of its own. A Boars Hill neighbor described it as follows:

> David's transport. It would bestow on it a dignity it perhaps shouldn't deserve to call it a motor car. It was what I think was called a Morris Minor Traveller, made in the Morris Car Works on the other side of the City from us—so perhaps a case of supporting local industry. It was of the sort of design which used to be called a Shooting Brake, though I hardly think the Lacks would have used it for its sporting purposes. It was extremely old, and none too reliable. His garage kept suggesting that 'Dr Lack might consider changing it for a better vehicle', but I cannot remember if he ever did while we were next door. Its unusual feature was that it had wooden supports on its bodywork, and in their case there was moss growing on the sills of these wooden battens, and the colour of the metalwork was—er—not quite as gleaming as it started out.[22]

It is perhaps little wonder that Bill Sladen, one of Lack's D.Phil. students in the mid-1950s, believed that his supervisor was jealous of his Ford Prefect.

The most important cause of conflict between Darwinism and Christianity concerns the nature of man. Christians believe that he has a spiritual nature ... while biologists hold that he has evolved from the beasts by natural selection. Are these views incompatible?
—*Evolutionary Theory and Christian Belief*, p 80

7

Evolutionary Theory and Christian Belief

A LETTER FROM John Wren-Lewis to David Lack inviting him to participate in a series of lectures at St. Anne's House, London, led ultimately to the publication of *Evolutionary Theory and Christian Belief, the Unresolved Conflict.* The series was entitled "Modern Cosmology and Christian Thought," and Wren-Lewis wrote:

> I am attempting in this course to get a group of people together to promote some honest thinking which neither assumes that science leaves no room for religion at all, nor, on the other hand, commits the dishonest act of saying that there is no longer any tension at all between the two fields, as so many religious people are willing to do nowadays on the strength of a few Christian declarations by scientists of note.[1]

The specific topic that David was to address in his lecture was "Man and Evolution—Modern Biological Science in Relation to Christian Ideas of Man's Place in Nature." From the contents of the letter, it is apparent that Wren-Lewis had first approached W. H. Thorpe of Cambridge University, who had demurred with the excuse that he was working on a book. Thorpe had suggested David as an alternative, stating that he had just completed his book (presumably *The Natural Regulation of Animal Numbers*).

The series of lectures was held in the autumn term of 1953 at St. Anne's House, and it comprised 10 lectures and dialogues culminating in a general discussion on "Theology and the Future of Science." Speakers, in addition to Lack, included the political philosopher Harry Burrows Acton, the theoretical physicist Charles Alfred

Coulson, the industrial economist Michael P. Fogarty, the English historian Cicely Veronica Wedgwood, and the physiologist Sir Russell Brain (later Lord Brain).

David presented his lecture, entitled "Evolution and Christian Belief,"[2] on November 26. The content and organization of the lecture bear strong similarities with those of *Evolutionary Theory and Christian Belief*. Lack sent a copy of the lecture to Peter Medawar, who was at the time Jodrell Professor of Zoology at University College, London. Medawar, who was knighted in 1965 and would share the Nobel Prize in Physiology or Medicine in 1980 with Frank M. Burnet for his work on tissue transplantation and the immune system, also contributed significantly to discussions of scientific method and the philosophy of science. In a letter dated January 18, 1954, Medawar urged David to publish his lecture.[3]

In the letter of submission to his editor at Methuen in September 1956, David titled his book, *The Conflict Between Darwinism and Christianity*, and stated that it was "not original, but rather a critical review."[4] He also asserted:

> My aim is to write fairly for both Christian and agnostic. Assuming you send the book to one or more readers, I would be most grateful to know of any places where I have been unfair to either. I am also most anxious to know whether it is possible, from reading the book, to know whether I myself am Christian or agnostic.[5]

He asked that royalties from the book be given to Magdalene College, Cambridge. Methuen published *Evolutionary Theory and Christian Belief, the Unresolved Conflict* in 1957.

Evolutionary Theory and Christian Belief (or "Daddy's Revolutionary Theory" as his children referred to it) began with a recounting of the famous confrontation between Thomas Henry Huxley and Samuel Wilberforce at a meeting of the British Association for the Advancement of Science in June 1860 (less than a year after the publication of Darwin's *Origin of Species*). The site of the confrontation was the University Museum in Oxford, the same museum where the swifts nest in the tower (see Chapter 6). No transcripts or minutes of the debate exist, and accounts of its content differ. Lack related, however, that Wilberforce, who at the time was Bishop of Oxford, after arguing against Darwin's theory in a polished and witty speech, concluded with a remark that was apparently intended to be humorous. The closing remark, variously recounted by different observers, was quoted by Lack as the recollection of Prof. Farrar, who was present at the meeting: "If anyone were to be willing to trace his descent through an ape as his *grandfather*, would he be willing to trace his descent similarly on the side of his *grandmother*?"[6] Less charitable versions of Wilberforce's concluding remark have been remembered, including one in which it is pointedly and derisively aimed at Huxley, inquiring on which side, his grandfather's or his grandmother's, he claimed ancestry from an ape. Huxley, who was to speak after Wilberforce, is reported to have whispered to a colleague, "The Lord hath delivered him into mine hands."[7] He then proceeded to present a straightforward and cogent description of Darwin's theory, concluding his presentation by saying that he would rather have a monkey for a grandfather than a man who used his great gifts to stifle truth.

Most of those present at the debate believed that Huxley had taken the day, but Lack noted that the controversy had only just begun. He identified four ways in which Darwinian theory was in conflict with mid-19th century Christian belief: (1) it contradicted the Genesis account of creation, thus calling into question the truth of the Bible; (2) it provided an alternative, naturalistic explanation for the principal philosophical argument for the existence of God, the argument from design; (3) it challenged the Christian concept of an historical Fall from a prior state of blessedness, asserting instead that man has arisen by natural selection from lower animals; and (4) it challenged the view that man was endowed with unique attributes by God (by being created in His image), again asserting that man's attributes are the product of evolution by natural selection. Lack described the positions of some of the major protagonists in the debate during the remainder of the 19th century, noting that there were scientists and churchmen on both sides of the issue. He also concluded that both the Roman Catholic Church and the Church of England had more or less come to an accommodation with the theory of evolution by the end of the century, with the salient exception of its implications for the origin of man.

The book continued with a description of the major tenets of evolution by natural selection as consolidated in the neo-Darwinian Synthesis of the mid-20th century. Lack's conclusions, summarized at the end of the book, were that "animal evolution is an historical fact," "man evolved from other animals," and "evolution is comprehensible in terms of the natural selection of hereditary variations [which] are random in relation to the needs of the animal, [and] the directions of evolution are determined by natural selection."[8] The last point, explored in detail in a chapter entitled "Chance or Plan?," addressed the misconception that because mutation is a random process, the products of evolution are due solely to chance. He rejected the idea that evolution has been directed by some creative "force," a position often taken by theologians attempting to reconcile Darwinism with Christianity.

Lack addressed questions about the nature of man in a chapter by that name, noting, as quoted in the epigraph above, that therein lies the major conflict between Darwinism and Christianity. He concluded that science has not yet accounted for the evolution of human attributes such as morality, truth, beauty, free will, individual responsibility, or self-awareness, all of which he considered to be central to the human experience. In a later chapter, he discussed the nascent field of evolutionary ethics and asserted that these efforts have proved largely unsatisfactory scientifically, as well as being displeasing to both the Christian and the secular humanist.

In his conclusions at the end of the book, Lack had suggestions for both the Christian and the scientist:

> All should accept the findings of science in the field of science.... On the other hand, it is important that the claims made by scientists in the name of science should relate to genuinely scientific matters, and that when they really refer to philosophical problems, this should be made clear. In particular, the claim that man has evolved wholly by natural means is philosophical and not scientific.[9]

One is left to wonder what Lack would have said about sociobiology (see Chapter 13). He concluded the book with a quotation from Alexander Pope's *Essay on Man:*

> Placed on this isthmus of a middle state,
> A being darkly wise and rudely great;
> He hangs between; in doubt to act or rest;
> In doubt to deem himself a god, or beast;
> Created half to rise, and half to fall;
> Great lord of all things, yet a prey to all;
> Sole judge of truth, in endless error hurled;
> The glory, jest and riddle of the world.

A second edition of *Evolutionary Theory and Christian Belief* was published by Methuen in 1961, and Routledge published a reprint of the first Methuen edition in October 2008 as part of its History and Philosophy of Science series. Lack added a chapter entitled "Afterthoughts" to the second edition, in which he addressed questions raised by works published after the release of the first edition, particularly the English translation of Teilhard de Chardin's *The Phenomenon of Man* (Harper-Collins, 1959). He rejected, on the same grounds that he rejected directed evolution, de Chardin's concept of a universal "psyche" in the animate world that allows for a progressive evolution culminating in the appearance of man. Lack concluded in "Afterthoughts":

> This suggests that the real gap is not between other animals and man, but between two methods of enquiry, scientific and philosophic, both of which are valid in the study of human nature, but only one of which, the scientific, is valid

in the study of other animals. If this is correct, it suggests on the one hand that no truly scientific theory can conflict with Christian beliefs, and on the other hand that the agnostic may accept the idea of man's evolution through natural selection without feeling that the basis of moral and other values is thereby undermined. . . . On any view, Christian or agnostic, a tremendous riddle remains.[10]

Evolutionary Theory and Christian Belief was widely reviewed in its first edition, including reviews in the popular press (i.e., *Telegraph* and *Times Literary Supplement*), in scientific and cultural journals (*Nature* and *Victorian Studies*), and in several religious publications. Except for a rather acerbic review from a biblical literalist (D. W. Wood) in *The Christian Graduate*, the appraisals were generally quite positive, with most reviewers concluding that Lack had accurately and lucidly depicted the fundamental conflicts between Christianity and evolutionary biology. He also apparently succeeded in masking his own beliefs quite effectively, because at least two of the reviewers declared that it was not possible to detect the author's personal beliefs.

Lack's personal view may have been best expressed in a paragraph at the end of an essay entitled "Natural Selection and Human Nature" that he apparently wrote sometime between the lecture at St. Anne's House and the publication of *Evolutionary Theory and Christian Belief*. Parts of the essay were incorporated in the latter work, but not this closing paragraph:

My personal view is that we must accept the scientific conclusion that man has evolved . . . by natural selection from (amoral) animals. But while this may appear to mean that man has no free-will, that he has no ethical (as opposed to social) behaviour, and that he has no reliable appreciation of truth, I accept the existence of these attributes in man because I consider that, although they come outside the limits of scientific investigation, and from their nature must always do so, yet they are valid, indeed essential, parts of the human experience. I admit that I cannot justify their validity in the same kind of way that scientific theories can be substantiated. I admit also that they seem irreconcilable with scientific knowledge. But both the findings of science and the moral experience seem far too valuable for either to be rejected.[11]

It was apparently to this declaration, that his Christian beliefs were irreconcilable with science and must so remain, that W. H. Thorpe was referring in his obituary for Lack, which was published in *Biographical Memoirs of Fellows of the Royal Society* (1974). Thorpe, a Quaker, was one who sought to reconcile Christianity and science in a seamless and coherent universal truth fully encompassing both world views. David instead seemed to hold a dualistic view, accepting the fundamental contradictions between the conceptions of man inherent in evolutionary biology and in Christianity but asserting that both have great value in our attempts to understand our lives and our place in the universe.

A final postscript to *Evolutionary Theory and Christian Belief* is provided by Lack's Huxley Memorial Lecture, "T. H. Huxley and the Nature of Man," which was presented at Birmingham University in 1962 and published in his *Enjoying Ornithology* (Methuen, 1965) (see Chapter 8). In the lecture, he focused on Huxley's principle of the "continuity of evolution," which suggests that human attributes such as free will, moral conduct, apprehension of truth and beauty, and self-awareness have rudimentary forms in prehuman animals. He discussed contemporary theories of evolutionary ethics, particularly those of C. H. Waddington and of T. H. Huxley's grandson, Julian Huxley, but concluded that none "of the advocates of evolutionary ethics explain the hardest problem of all, of why man, knowing what is good, so often chooses the evil."[12]

The publication of *Evolutionary Theory and Christian Belief* also led to two BBC Radio appearances by Lack in the late 1950s. The first, broadcast on June 7, 1958, was entitled "Religion and Evolution" and took the form of a debate between Lack and David Newth, an embryologist at University College, London. Peter Medawar served as the moderator for the debate. The second broadcast was part of a series entitled *Religion and Philosophy: Guidance in a Scientific Era*, in which Lack represented the views of the modern biologist. It was broadcast on October 6, 1959.

Lack also delivered a lecture to the Friday Evening Discourse of the Royal Institution on May 18, 1960. Initiated informally in 1825 by Michael Faraday, the Discourses became a significant window into the scientific enterprise for the educated public. The substance of Lack's presentation was subsequently published in *Nature*.[13]

PERSONAL RELIGIOUS HISTORY

David's parents belonged to the Church of England, and the children were all baptized by the church, but their home was essentially irreligious. With his strong scientific interests, particularly in evolution, David quickly adopted an agnostic position, which he retained through his schooling and university years, while teaching at Dartington Hall, and during the war. However, his interest in choral music, which had begun at Magdalene College, meant that he was at times during those years regularly attending Anglican services and singing in the choir.

He converted to Christianity in 1948, attributing his change in belief to continued interactions with friends from Dartington Hall, whom Elizabeth Lack later identified as Dan and Mary Neylan. The Neylans were both teachers at Dartington Hall, and they were two of David's closest friends while he taught there. He was particularly fond of their daughter, Sarah, who was a young child during his latter years at the school, and the Neylans asked him to be godfather of their second daughter, Elizabeth, a favor that the Lacks later returned when they asked Dan to be the godfather of their second son, Andrew. The Neylans remained close friends even after David left Totnes.

Others have suggested that it was Elizabeth herself who had a significant influence in David's decision to convert. When I asked Elizabeth to give me three descriptors of David, her second was that he was "a definite Christian."[14] When she arrived in Oxford in 1946, Elizabeth visited numerous churches before deciding to attend St. Giles. She took David there for the first time sometime before they were married, thus introducing him to St. Giles' vicar, Canon Reggie Diggle.

Upon his conversion, David attended confirmation classes with Canon Diggle, whom one family member described as "a remarkable man, one of the most naturally spiritual people I have ever met."[15] David and Elizabeth attended St. Giles after their marriage in 1949, and they continued to attend there even after moving to Boars Hill. David served both as a regular reader and as a sideman (usher). Canon Diggle baptized all four of the Lack children. David and Elizabeth began attending their local parish church, St. Peter's Church in Wootton, only after Canon Diggle retired from St. Giles.

FIGURE 14: Catherine Lack's christening day, in the garden at Hatherton, 1959. From left: Godmothers Christine Pardoe and Joyce McMaster, godfather Chris Wilson, Elizabeth holding Catherine, Canon Diggle, and David. (Photograph courtesy of Lack family.)

Few hints regarding other influences in his decision to convert are evident in David's writings. Near the end of the lecture at St. Anne's, he referred to C. S. Lewis, a professor of Medieval and Renaissance Literature and a major Christian apologist of the mid-20th century, in the following passage:

> Even omitting the peculiar qualities of man, I see no justification for the view that science is capable of providing a complete account of the world. To cite C. S. Lewis's analogy, why should not biology bear to nature a relationship similar to that which the material analysis of a poem bears to the poem itself.[16]

Lewis was a tutor at Magdalen College, Oxford, when David was elected a member of the Senior Common Room in January 1947. Although David was acquainted with Lewis, the two men had markedly different personalities and world views, and it is unlikely that there was much direct contact between them. They did not get on well, a fact that rather surprised David. David was not a frequent participant in the Magdalen Senior Common Room, which he found very "clubby." The current archivist stated that the archives suggest only that "Lack is distinguished by his absence."[17] Andrew remembered, however, that his father preferred Lewis's *Space Trilogy* to J. R. R. Tolkien's *Lord of the Rings*, suggesting that "perhaps he had more sympathy with the Protestant perspective than with the Catholic one."[18]

Other personal influences on David's faith, besides Canon Diggle, included Elizabeth's brother-in-law, Michael Wilson, and Peter Hartley. Wilson, who was the nephew of Dr. Edward Wilson, the physician on the ill-fated Scott expedition to the South Pole in 1912, was himself a physician. He was ordained a deacon in the Anglican Church in 1953 and, after returning from medical work in the Gold Coast (now Ghana) in 1955, became curate of St. Mary's Church, Leigh, Lancashire. In 1958, he became Chaplain of the Guild of Health, Edward Wilson House, London (named for his late uncle) and an assistant at St. Martin in the Fields. Hartley, who had an M.A. degree from Queen's College, served as senior research officer at the Edward Grey Institute of Field Ornithology from February 1947 until January 1951, when he left to become warden of Flatford Mill Field Centre in Suffolk. He subsequently took Holy Orders and became rector of St. John the Baptist Church in Badingham, Suffolk. David and Elizabeth chose Hartley to be the godfather of their first child, Peter.

After David received his appointment to Trinity College, Oxford, in 1963 (see Chapter 8), he began attending evening prayers on most Sundays during term. He established a friendship with the chaplain at Trinity, Leslie Houlden. Houlden described him as a very sweet, unassuming person, who, along with the chemist, James Lambert, was one of the supportive Professorial Fellows. When Houlden left

Trinity in 1970 to become Principal of Cuddeedon College (now Ripon College Cuddeedon), Trevor Williams replaced him. Although Williams did not develop a close relationship with David, he did visit him regularly during his terminal illness.

Not unlike many post-Enlightenment Christians before him, David had difficulty understanding the place of miracles in a scientifically informed world view. After he and Elizabeth read Ruth Cranston's *The Mystery of Lourdes* (Evans Bros., 1956), he wrote lengthy inquiries to Michael Wilson (then chaplain at the Edward Wilson House) and Fr. Lawrence Bright of Hawksyard Priory in Rugeley, Staffordshire. With his usual logical approach to a problem, he posed the following alternatives to the pair:

1. Do you personally consider that the Lourdes cures are established beyond reasonable doubt? ...
2. Do you think an unbiased medical commission likely to accept them ...?
3. On the view that the spectacular cures at Lourdes are not possible of explanation, now or in the future in medical terms, there would seem to be three types of explanation: (a) that the mind has powers over the body hitherto undreamt of. ... (b) that the Lourdes cures simply illustrate, if with rather extreme examples, the general efficacy of prayer and other spiritual factors in healing the body, ... (c) while not denying that (b) may be occurring at Lourdes, the spectacular cures are to be regarded as miracles in the usual Christian meaning of the term.[19]

Both men responded to Lack's queries with lengthy and theologically based answers, but there is no indication of Lack's own views on the question of miracles.

CHRISTIANITY AND THE LACK FAMILY

David took an active role in planning the spiritual development of his children. When Peter, his first child, was not yet 4, David wrote to Peter Hartley inquiring about introducing children to prayer and when one should begin taking them to church. Hartley's reply congratulated him on beginning to teach prayer at a young age and suggested that the prayers should express gratitude and be "immensely concrete." With regard to church attendance, his suggestion was that the Lacks should begin to take their son to church at once, preferably Mattins or Evensong, and "let him stay the whole service," and encourage him to sing. He also enjoined, "ABOVE ALL, DO NOT LET HIM GO NEAR CHILDREN'S SERVICES, SUNDAY SCHOOLS, JUNIOR CHURCHES ETC. ETC."[20]

When Peter and Andrew were 5 and 3, respectively, David and Elizabeth responded to a church program entitled "Operation Firm Faith"—the literature for which expressed the hope that the family would go to church together and showed a father, mother, and two children doing so—by writing letters soliciting advice from Michael Wilson, then at St. Mary's in Leigh, and Peter Hartley, in Badingham. In typical Lack fashion, they prepared a summary of their conclusions, "Some Letters on the Family at Church," that both expresses frustration about their inability to attend as a family and provides some interesting insights into the Lack family of the mid-1950s. One section declared:

> Our parents could take their children to church with them because (i) they normally went at 11 a.m. not 8 a.m., and (ii) they had a cook to prepare Sunday dinner. . . . The issue in (ii) is not trivial, because we think Sunday dinner a valuable family occasion, the children eat better then than in the week, and anyway will have school lunches on weekdays, we live two miles from any stores and the butcher delivers so late on Saturday that a Sunday joint is almost inevitable. So the mother cannot attend church at 11 a.m. as she must cook, and either mother or father must be at home at 6 p.m. to put the children to bed.[21]

The paper went on to explain that this situation would last for several more years, meaning that the family would be unable to attend church together on a regular basis for a long time. Indeed, it was not until Peter and Andrew were attending Dragon School in North Oxford that the family began to attend the services there. Until that time, David and Elizabeth continued to alternate attending the 8 a.m. and 11 a.m. services at St. Peter's Church in Wootton alone.

The children recalled that their parents taught them to say the Lord's Prayer and to pray at bedtime. Other than that, they remembered no other religious activities at home.

David continued to attend services regularly throughout his life. After his death on March 12, 1973, a funeral service was held at St. Peter's Church, Wootton, the Rev. Gordon Owen presiding (see Chapter 12). His ashes were buried in the churchyard cemetery there. Elizabeth continued to attend St. Peter's regularly until the infirmities of age prevented her from attending services.

If you decide to take up a particular species, and this is one of the most enjoyable of all studies, you must love your bird.

—*Enjoying Ornithology*, p 20

8

Enjoying Ornithology

"MAD KEEN ON birds." This was the first description of David Lack that Peggy Varley, widow of his colleague and friend George Varley, gave when I asked her for three descriptors of Lack.[1] And indeed he was persistently and passionately interested in birds of all kinds, at all times, and in every place he visited. One of his sons commented that "he was always more interested in birds than in people,"[2] and another, as noted in Chapter 6, said that it was not until he left home to go to Bryanston that he realized that not every family spent their holidays going somewhere to observe "beastly birds."[3]

Enjoying Ornithology, published by Methuen in 1965, was intended to convey that enthusiasm for studying birds to bird watchers and amateur ornithologists. It is a collection of some of Lack's published papers; articles from popular scientific magazines, newspapers, and an anthology; radio broadcasts; and addresses presented to various audiences. The book was organized into five sections: Migration, Interlude, Some British Ornithologists, Darwinian Evolution, and Entertainments.

More than half of the book was devoted to the Migration section, which chronicled some of the key findings from Lack's long interest in the topic. In particular, it included reprints of some of the papers describing the results of radar studies of migration carried out at an R.A.F. radar installation in Norfolk by Lack and other personnel from the Edward Grey Institute of Field Ornithology (EGI) (discussed later). In contrast to much of his other work, which focused on the adaptive significance of ecological, behavioral, or morphological traits, Lack's migration studies were primarily descriptive, and he used his findings to draw inferences about the mechanisms underlying bird migration.

The studies revealed that birds migrating over the North Sea to and from the Low Countries or Scandinavia typically fly at altitudes between 3000 and 6000 feet, and frequently much higher—altitudes at which they could not be observed visually from the ground. By comparing the radar records of movements with those made simultaneously by direct visual observation from the ground, Lack concluded that the visual records often conflicted with those detected by radar, and this frequently resulted in misleading conclusions about the direction of flight of the large majority of birds. The radar records also provided evidence of large-scale temporary movements of both shorebirds and small songbirds away from the path of approaching storms, known as weather-related movements, something Lack had previously observed in the common swift (see Chapter 6). Lack also noted that migrating birds became disoriented when they encountered a cloud cover that obscured the open sky, which helped to confirm the results of laboratory studies on migratory orientation suggesting that daytime and nocturnal migrants use solar and stellar cues, respectively, as their primary means of orientation.

Probably his most important observation, however, was related to the effect of crosswinds on migrating birds. The radar records showed that in the fall both shorebirds and songbirds tended to leave on their trans–North Sea migration with a following wind, establishing an appropriate compass orientation to reach their destination. However, they often encountered crosswinds during the flight. In this event, the birds would maintain their established compass direction and would therefore drift off-course somewhat, depending on the direction and speed of the wind. Lack concluded that the birds did not compensate for changes in wind direction during flight, but instead displayed compensatory changes in direction after making landfall.

The Entertainments section concluded with reprints of two tongue-in-cheek papers that originally appeared in *Bird Notes*, a journal published by the Royal Society for the Protection of Birds. The first, entitled "A Vision of Rome, 1960" (Rome was the site of the XVIII Olympiad), was published in 1959 by "Cassandra Lark" (aka David Lack). It was purportedly based on the author's observations at the 1958 International Ornithological Congress (IOC) in Helsinki, at which "Britain was represented by five ornithologists and a hundred bird-watchers (An ornithologist: one who attends lectures illustrated by graphs; a bird-watcher: one who slips away from such lectures in order to add a new species to his life-list; *The New Ornithological Dictionary*.)"[4] The essay laid out a vision of the first "Bird-watching Olympic" competition, in which national teams of amateur bird watchers compete to see as many bird species as possible between dawn and dusk. The teams enter the stadium in a predawn ceremony accompanied by appropriate national music such as the *Thin-billed Nutcracker Suite* for the Russians. One British contestant is unfortunately disqualified ("he once used a histogram"[5]). As the teams return with their lists, the band quietly plays the "Catalogue Song from *Don Giovanni* (spiritual, if not actual, ancestor of all life-listers)."[6] Many British bird watchers, including his friend from Gresham's School, Maury Meiklejohn ("Professor Hoodwinksbane") and his close friend and former colleague in the EGI, Peter Hartley ("Peter Pugnax"), appeared in the essay.

The second tongue-in-cheek paper was published in 1963 and marked the retirement of Alister Hardy as Head of the Department of Zoological Field Studies at Oxford (discussed later). The paper was entitled, "An Undiscovered Species of Swift" and purported to describe a new swift species, the Hardy's swift (*Apus durus*). Lack described the biology of Hardy's swift and asserted that it feeds on the abundant "aeroplankton" above 1000 m, thereby avoiding competition with the common swift, which feeds lower. The EGI received numerous inquiries from people wishing to add the new species to their life list.

Enjoying Ornithology was illustrated by Robert Gillmor, marking the beginning of a collaboration between Lack and the noted bird artist that greatly enhanced some of Lack's later books, notably *Population Studies of Birds, Ecological Adaptations for Breeding in Birds, Ecological Isolation in Birds*, and *Evolution Illustrated by Waterfowl.*

EGI PERSONNEL

As noted at the conclusion of Chapter 5, the year 1955 saw the second retirement of W. B. Alexander, the EGI's first Director and subsequently its volunteer Librarian. Also by that date, several of the first crop of EGI doctoral students had completed their degrees and left the institute. At the beginning of 1956, only three remained: David Snow as a senior research officer, and John Gibb and Monica Betts, both research officers assigned to the Breckland Research Unit (see Chapter 5). Snow and Gibb left the institute in 1957, and Betts in 1958. Reg Moreau remained in his half-time research post and also continued to serve as editor of *Ibis*. An Oxford undergraduate who had been appointed as a quarter-time field assistant in 1953, Denis Owen, continued in that position at the beginning of 1956.

Two new doctoral students arrived in 1955: Bernard Stonehouse in April, to write his thesis after his return from South Georgia, and Doug Dorward in October. Two other D.Phil. students arrived in September 1957: Chris Perrins, who was to have a profound influence on the future of the EGI (see Chapter 10), and Philip Ashmole. Stonehouse, Dorward, and Ashmole all participated in the Ascension Island expedition (discussed later), and Perrins worked primarily on the long-term tit study in Wytham Woods.

FIGURE 15: EGI personnel, 1958. Seated from left: R. J. Andrew, Reg Moreau, David, Monica (Betts) Turner, D. F. Owen. Standing from left: Bill Bourne, Chris Perrins, Mary Goodacre (librarian), Robert MacArthur, W. L. N. Tickell. (Photograph courtesy of the Edward Grey Institute of Field Ornithology.)

Two postdoctoral students arrived in the EGI during 1957–1958. The first, Robert H. MacArthur, who had just completed his Ph.D. at Yale under the supervision of G. Evelyn Hutchinson, arrived in October 1957. The second, Richard J. Andrew, arrived in February 1958 after completing his Ph.D. under Robert Hinde at the Madingley Field Station at Cambridge. MacArthur, who was to become a pioneering theoretical ecologist (see Chapter 11), studied the ratio of migrant to resident bird species in various British habitats. Andrew, whose interests were more behavioral than ecological, studied the mechanisms behind mobbing behavior in blackbirds.

The early 1960s saw the arrival of several of Lack's D.Phil. students. David Seel joined the institute as a research student in October 1960 and studied the breeding biology of house and tree sparrows in Wytham Woods and at the nearby University Farm. Roger Bailey and Ian Newton arrived in October 1962, Bailey as a field assistant and Newton as a research student. Bailey spent 18 months as marine ornithologist aboard the *Discovery III*, the main British ship participating in an international exploration of the Indian Ocean. Newton remained closer to home, studying the food of several finch species, particularly the bullfinch. The Agricultural Research Council provided funding for his research because of the economic impacts of the bullfinch's depredation of fruit buds in orchards. Tom Royama became the first international D.Phil. student when he arrived from Japan in October 1963 with funding from the British Council. The following October, Dr. Peter Evans joined the institute after completing a Ph.D. in chemistry at Cambridge, and in October 1965, a second international student, Uriel Safriel of Israel, arrived, also supported by the British Council. Royama joined the team working on the tits in Wytham Woods, pioneering the use of back-mounted cameras to study the food fed to nestlings, while Evans joined the team working on migration. Safriel worked on the oystercatcher on Skokholm, an uninhabited island lying 4 km off the coast of Pembrokeshire, Wales. The arrival at the EGI of international doctoral students underscored the growing reputation of the institute.

A major administrative change in the Department of Zoology had ramifications for the EGI. In 1961 Sir Alister Hardy, who had been appointed as Linacre Professor of Zoology and Head of the Department of Zoology shortly before Lack's selection as Director of the EGI, retired from those posts 2 years early to make way for the appointment of J. W. S. Pringle as Linacre Professor and department head. The early appointment of Pringle was made to enable him to begin oversight of the construction of a new home for the Department of Zoology (see Chapter 12). Hardy continued as Head of the Department of Zoological Field Studies, comprising the EGI and the BAP, until his mandatory retirement in July 1963. Pringle appointed David Lack to fill that position upon Hardy's retirement. The meeting at which Lack's appointment was announced culminated in a rare public outburst by the normally diffident Charles Elton, who believed that his seniority was being overlooked (see Chapter 11).

No doubt the saddest and most tragic time in Lack's 28 years as Director of the EGI occurred in April 1962. Stephen Lee had joined the EGI in September 1961 as a research officer to work on the migration research (described later) after completing a degree in physics at Cambridge. He was monitoring visual seabird migration from the cliffs of Lewis (Scotland) on April 21, 1962, when he fell to his death at the age of 22. Lack wrote:

> [H]e had already become the friend of all of us here, as he had an extremely attractive personality. He was also a very capable observer of migration and was showing considerable originality in his approach to his research. . . . He showed such promise that he would surely have become an important figure in British ornithology.[7]

Another sad though not tragic day came in August 1964, when Reg Moreau retired after 17 years as half-time research officer in the EGI. In this position and as

editor of *Ibis*, Moreau had become a significant figure in 20th century British ornithology, and he was later recognized, along with David, as a major contributor to the
transformation of what constituted scientific ornithology in Britain.[8] He was not
only Lack's valued colleague in the EGI but also, since David's trip to Tanganyika in
1934 (see Chapter 2), a close friend.

ASCENSION ISLAND EXPEDITION

The British Ornithologists' Union (BOU) was founded in 1858 to promote the
scientific study of birds in Great Britain. It inaugurated publication of its principal
journal, *Ibis*, the following year. In preparation for the celebration of its centenary,
the BOU sponsored an expedition to study the seabirds of Ascension Island, a
volcanic island in the mid-Atlantic approximately 8 degrees south of the equator.
The BOU committee selected Bernard Stonehouse, who had recently completed
his D.Phil. thesis on the breeding behavior of penguins on South Georgia under
Lack's supervision (see Chapter 10), to head the expedition. Two research students in the EGI, Doug Dorward and Philip Ashmole, were also chosen to participate. In addition, Sally Stonehouse, Bernard's wife, accompanied the three
scientists and served as her husband's field assistant. The party left in the fall of
1957 to spend 18 months on the island; they also visited the island of Saint Helena,
another volcanic outcrop lying more than 1500 km southeast of Ascension Island,
before returning to Oxford.

Stonehouse devoted his attention to the breeding behavior of the Ascension
Island frigatebird and also studied two species of tropicbirds. Ashmole studied the
breeding cycle of wideawake terns, and Dorward studied two species of boobies
breeding on the island, both writing up their research to complete their doctorates
under Lack's supervision (see Chapter 10).

RADAR STUDIES OF MIGRATION

As the content of *Enjoying Ornithology* indicates, David Lack's intense interest in
migratory movements of birds continued, particularly during the first few years of
the 1956–1966 period. This interest was further stimulated in the fall of 1957, when
the Research Branch of the Fighter Command of the R.A.F. invited him to use some
of their radar installations to study the radar "angels." The invitation spurred the
development of a major research program in the EGI involving several personnel,
particularly after the institute received funding from the Department of Scientific
and Industrial Research (DSIR) in 1958. DSIR funding continued until 1964.

The research program involved the use of R.A.F. radar stations in northeastern Scotland and Norfolk that were monitoring movements of birds, particularly over the North Sea, and a comparison of the radar records with visual observations made on the ground. Several bird observatories cooperated by providing concurrent records of birds they had observed. These included, at various times over the next few years, observatories at Cley, Dungeness, Gibraltar Point, and Spurn, and on Fair Isle and the Isle of May. Detailed meteorological data were also collected to determine the effects of wind speed and direction, cloud cover, and frontal systems on bird movements. In 1961, the EGI program began a cooperation with a similar study being done by a research team at Marconi's Wireless Telegraphic Laboratories in Great Baddow, Essex.

Several EGI personnel in addition to Lack were involved in the radar studies, including two D.Phil. students, Bill Bourne and Peter Evans, and several research officers including M. T. Myres, J. L. F. Parslow, S. L. B. Lee, J. Wilcock, and C. M. Reynolds. Most of the research officers were hired specifically for the radar study and were supported by the DSIR grant. As noted earlier, one of these researchers, Stephen Lee, tragically lost his life while working on the project.

Many of the most important findings from the radar studies were described in *Enjoying Ornithology* (discussed earlier). Lack also presented some of the conclusions of the research in a paper entitled "Weather Factors Initiating Migration" at the XIII IOC in Ithaca, New York, in June 1962.

STUDENT CONFERENCES

Eleven Student Conferences in Bird Biology (numbers 10–20) were held at the EGI between January 1956 and January 1966. The topics ranged across several subdisciplines of avian biology: anatomy and physiology (Bird Flight, 1957), behavior (Song, Fighting and Territory, 1956; Behaviour, 1961; Migration, 1959; Orientation, 1962), ecology (Population Problems, 1958; Tropical Ecology, 1960; Population Dynamics, 1963; Predation, 1964), and evolution (Adaptive Radiation, 1965; Sexual Selection, 1966). Student attendance at the conferences increased considerably over the period, ranging from 30 attendees in 1956 to more than 50 in 1965 and 1966.

As in the earlier conferences, all papers presented were by students except for those of invited senior visitors addressing the main topic of the conference. Senior visitors included such prominent scientists as Lawrence Brower, A. J. Cain (three times), Dennis Chitty, John Crook, Robert Hinde (twice), Hans Kluyver, Peter Marler, John Maynard-Smith, G. V. T. Matthews, C. J. Pennycuick, A. C. Perdeck, Frank Pitelka, W. H. Thorpe, Niko Tinbergen (four times), and V. C. Wynne-Edwards.

The last day of the 1963 conference on Population Dynamics featured a special lecture by Wynne-Edwards, whose book, *Animal Dispersion in Relation to Social Behaviour*, had been published the previous year. The discussion following Wynne-Edwards' presentation was led by four respondents, each of whom began with a short presentation related to Wynne-Edwards' theory of intrinsic population regulation (see Chapter 9). Lack described the discussion as follows:

> The discussion lasted from 11 a.m. to 1 p.m., and again from 2 p.m. to 4:30 p.m., and there is no question that it could have been continued much longer. This shows the great interest the subject aroused, and a special debt is due to Professor V. C. Wynne-Edwards for his clear exposition and for the sustained, but always friendly, arguments between him, the main discussants and a large number of other members of the conference. It was the best public scientific discussion on any subject that most of us had ever heard.[9]

The discussion was indeed to continue, as Lack's opposition to Wynne-Edwards' ideas on the role of group selection in population regulation prompted him to write a book in response. Their debate is the principal subject of Chapter 9.

BROOD-SIZE MANIPULATION STUDIES

Lack's theory that clutch size in birds that feed their young is selected to maximize the reproductive output of the individual or the pair, and that the ability of the parents to provide food to the developing young is usually the limiting factor (see Chapter 3), stimulated many studies attempting to test its predictions. The most common type of study involved comparing the average number of young fledging from broods of different sizes. The principal prediction of Lack's theory was that brood sizes corresponding with the normal clutch size in the population would be the most productive. Larger-than-normal broods would actually result in fewer fledged young per nesting attempt than normal broods.

In the spring of 1960, Chris Perrins, then a field assistant in the EGI, initiated brood-size manipulation studies of great and blue tits in Wytham Woods and of the common swifts nesting in the museum tower (see Chapter 6). By transferring hatchlings among nests at the time of hatching, he created both smaller- and larger-than-normal broods. The results for the tits were inconclusive, with the normal-sized broods being the most productive in some years, whereas in other years the larger-than-normal broods were more productive. The results for the common swift were generally consistent with Lack's hypothesis.

Brood-size manipulation was subsequently used by other researchers from the EGI. Ian Newton, one of Lack's D.Phil. students, used it in his studies of the bullfinch, and J. B. Nelson, who was a research student in the EGI in 1959–1960 but left to study with Niko Tinbergen, performed twinning experiments in gannets, which normally have a clutch size of 1. The results were inconclusive, ultimately leading to modifications and extensions of Lack's original hypothesis (see Chapter 9).

TRINITY COLLEGE

Trinity College, Oxford, was founded in 1555 by Sir Thomas Pope, a Catholic lawyer who had risen to prominence as a trusted counselor to Henry VIII. In that year, Pope was a privy counselor for Mary Tudor, Henry's eldest daughter, who had succeeded her brother Edward VI in 1553. She is remembered as "Bloody Mary" for her violent enforcement of the return of England to Roman Catholicism, and two of the most notorious acts of violence took place literally within a stone's throw of the entrance to the newly chartered Trinity College. In October 1555, Nicholas Ridley, Bishop of London, and Hugh Latimer, Bishop of Worcester, were burned at the stake in the middle of Broad Street outside the entrance to Balliol College, which stands next door to Trinity. Five months later, the Archbishop of Canterbury, Thomas Cramner, suffered the same fate. All three had been found guilty of heresy after being interrogated by Spanish friars. An iron cross sunk in the bricks of Broad Street still marks the site of their executions.

Pope's endowment of Trinity was to support 12 Fellows, 12 scholars, and up to 20 undergraduates, all men. All Fellows were required to take Holy Orders and to remain celibate. The number of Fellows remained unchanged until 1939, when Trinity began to add additional Fellows, and women were accepted beginning in 1979. Today the college has about 40 Fellows, several of whom are women, and about 30 graduate and 200 undergraduate students. Some of its most distinguished alumni include William Pitt (the Elder), well-known English statesman of the mid-18th century; John Henry Newman (Cardinal Newman), a major religious thinker and educational philosopher in the 19th century; and Sir Terence Rattigan, one of England's most renowned dramatists of the 20th century. The promising young atomic physicist, Henry Moseley (creator of Moseley's law), who was killed at the Battle of Gallipoli at the age of 27, was also an alumnus of Trinity.

David was elected a Professorial Fellow of Trinity in 1963. Unlike his experience as a member of Magdalen College, Oxford, which he found too "clubby" and where he was "distinguished by his absence" (see Chapter 7), David truly enjoyed his affiliation with Trinity. He frequently lunched at the college, and he often attended dinners there on Thursday evenings before the weekly college meetings. He appreciated the intimacy of the small college, and he developed a friendship with Hans Krebs, a bio-chemist and winner of the Nobel Prize in Physiology or Medicine (1963), who was also a Fellow of Trinity. He also greatly appreciated the Chaplain, the Rev. Leslie Houlden. David's obituary in the *Trinity College Record* described the service at which he was admitted to fellowship:

> When he came into the chapel to be admitted to his fellowship he was de-lighted to hear David Raven playing some of Papaeano's music, and in his speech on Trinity Monday (old style) he spoke of this and of his happiness that he was no longer having to await the appearance of his Papagena.[10]

XIV INTERNATIONAL ORNITHOLOGICAL CONGRESS

At the XIII IOC, held in Ithaca, New York, David Lack was elected president of the XIV IOC, to be held 4 years later somewhere in the United Kingdom. One of his first tasks as president was to locate an appropriate site for the Congress, and his preference was to hold it somewhere in Scotland, with a pre-Congress excursion in the same area. Sites under consideration included Edinburgh Uni-versity, Saint Andrews University, and the University of Aberdeen, as well as Norwich, county seat of Norfolk and home of the University of East Anglia. But after much deliberation, Oxford, which had hosted the IOC in 1934, was chosen to host again.

Niko Tinbergen served as secretary of the Congress, and David praised him for the quantity, diversity, and quality of work that he did to make the Congress a suc-cess. David's presidential address, entitled "Interrelationships in Breeding Adapta-tions as Shown by Marine Birds," foreshadowed his soon to be published *Ecological Adaptations for Breeding in Birds* (see Chapter 10), and the opening line of the address could also summarize the conclusions in that book:

> The theme of this address is that the many adaptations for breeding in birds cannot be evaluated separately, but are closely interrelated with each other and with ecological factors such as food and predation.[11]

As part of the Congress, Lack called a meeting to discuss a proposal that he put forward, to appoint a small committee to establish a uniform taxonomic order for the birds of the world that would be accepted across international lines and used in all major ornithological journals. The call for the meeting asked that individuals who wished to address the issue submit their names to the Congress Secretariat. The meeting generated considerable heat but little light. As one commentator expressed it, those who wished to support the proposal but not to speak were invited to submit their "assent without qualification," but that there was no option to "signify *dissent* without qualification."[12] No decision was reached.

There is no evidence for the view of Wynne-Edwards (1955, 1962) that such deferred maturity has been evolved through group-selection in long-lived species to reduce the number of young and so prevent over-population.

—*Population Studies of Birds*, p 275

9

Population Studies of Birds

CHARLES DARWIN CHARACTERIZED what ecologists today would refer to as the *biological community* or the *biocenosis* as "an entangled bank"—an array of plant and animal species that interact in complex ways to form the self-sustaining communities that the naturalist observes. A significant proportion of the 20th-century focus of ecology was devoted to attempts to understand the web of relationships among plants, animals and microorganisms and with the nonliving components of their environment that maintain these communities. In *An Entangled Bank: The Origins of Ecosystem Ecology*,[1] Joel Hagen used Darwin's term not only as a descriptor of the complexity of the biological community but also as a metaphor for the web of conflicting ideas that have been developed by ecologists attempting to understand the workings of biological communities. One of the overarching ideas about which much controversy centered was the "balance of nature" concept, an idea born in part of the apparent temporal stability of many biological communities.

David Lack entered this fray with the publication of *The Natural Regulation of Animal Numbers* in 1954 (see Chapter 5). Ironically, another book attempting to explain the determination of population numbers, by two Australian entomologists, H. G. Andrewartha and L. C. Birch, was published that same year.[2] As described in Chapter 5, Lack's book emphasized the relative stability of many populations over time, at least compared to their inherent ability to increase exponentially in size, and concluded that populations are controlled by density-dependent mortality factors, sometimes by predators but usually by food availability. This conclusion was in conformity with the balance of nature concept. Andrewartha and Birch, on the other hand, emphasized the occurrence of large and variable fluctuations in population size and concluded that population numbers are usually determined by density-independent factors, often by climatic variation. They explicitly rejected the balance of nature concept.

Another challenge to Lack's conclusions came from V. C. Wynne-Edwards, who had written a review critical of *The Natural Regulation of Animal Numbers* in *Discovery* (see Chapter 5). Although Wynne-Edwards concurred with Lack that food was the principal factor setting ultimate limits on most populations and that most populations are relatively stable over time, he proposed a radically different mechanism for the achievement of this balance in nature. He presented his alternative hypothesis for the regulation of animal numbers in an important book, *Animal Dispersion in Relation to Social Behaviour* (Edinburgh, 1962).

The publication of *Animal Dispersion in Relation to Social Behaviour* served as the immediate stimulus prompting Lack to write *Population Studies of Birds*, a book that might well be subtitled, "A Defense of *The Natural Regulation of Animal Numbers*." Lack originally intended to include this defense as a major part of the text, but ultimately he changed his mind and included it only as an extensive (33-page) Appendix entitled, "The Theoretical Controversies Concerning Animal Populations."

The body of *Population Studies of Birds* comprised descriptions of the principal findings from population studies of bird species that had "continued for at least four years and . . . consist[ed] of more than just an annual census."[3] Studies appearing since the publication of Lack's earlier book that met these criteria involved 12 species; one study of another species, the quelea of sub-Saharan Africa, was also included despite not meeting the 4-year requirement because it was the only study of a species from a seasonal tropical environment. In addition to the 13 primary species covered, brief discussions of shorter studies of 12 other species were included in chapters where there were parallels with the major species treated. There was considerable taxonomic, geographical, and habitat diversity among the species treated, a fact that Lack suggested gave robustness to his conclusions regarding population regulation in birds.

Lack's conclusions closely paralleled those in his earlier book. He continued to assert that clutch size in birds was selected to maximize the reproductive output of the individual and that density-dependent mortality was the primary means of regulating bird populations. He also suggested that starvation-related mortality due to shortages in the quantity or quality of food was the principal mechanism involved. One topic on which the new evidence required a revision of his earlier conclusions was related to the timing of the breeding season in a species. Lack had earlier suggested that egg laying should occur at a time that ensures that hatching will coincide with the maximum availability of food for chicks. However, from the results of several of the newer studies, he concluded that the timing of egg laying is constrained by food availability for egg production by the female, resulting in hatching at a later time than would be optimal.

The Appendix was divided into three sections. The first, entitled "Density-Dependent Regulation," consisted of short summaries of the 24 chapters of *The Natural Regulation of Animal Numbers* and closed with the identification of four major challenges to Lack's conclusions that had emerged since its publication. Two of the challenges came from Andrewartha and Birch's work: (1) skepticism about Lack's conclusion that density-dependent factors operate to control population size, and (2) rejection of the competitive exclusion principle and the role of interspecific competition in affecting the distribution and abundance of animal populations. As noted earlier, Andrewartha and Birch concluded that density-independent factors were primarily responsible for controlling animal populations, and they asserted that natural populations rarely achieve densities great enough for either intraspecific or interspecific competition for resources to significantly impact population size.

The other two challenges to Lack's conclusions came primarily from V. C. Wynne-Edwards, who rejected the views that (1) food shortages resulting in starvation-related mortality are controlling most populations and (2) individuals are selected to maximize their individual reproductive rate. As noted earlier, Wynne-Edwards' alternative conclusions were expressed most thoroughly in his book, *Animal Dispersion in Relation to Social Behaviour*. His alternative theory of population regulation required that natural selection operating at the level of the population (i.e., group selection) would override selection at the level of the individual, which favors maximization of the individual reproductive rate.

Lack addressed these two sets of challenges in the two remaining sections of the Appendix, "The Attack on Density-Dependence" and "Animal Dispersion." He frequently referred to the results of the studies covered in the main body of the book to buttress his own views contradicting those of his critics. Although he was equally dismissive of both sets of challenges, he reserved his most withering criticisms for Wynne-Edwards' views, particularly his group selection hypothesis.

Reviews of *Population Studies of Birds* were generally favorable, but one stands out by contrast with the others, and indeed it is probably one of the most memorable reviews ever published in the flagship journal of the Ecological Society of America, *Ecology*. Dennis Chitty, who had spent almost 25 years at the Bureau of Animal Population and earned a D.Phil. under Charles Elton's supervision before moving to the University of British Columbia, wrote the review, which began as follows:

> Population ecologists spend much of their time trying to find out why birth-rates and death-rates tend to equal one another in the long run in spite of being annoyingly variable from year to year. The relevant observations are usually

imprecise and consistent with several alternative explanations, which in turn are as plausible as they are difficult to test. It is not surprising therefore that the supporters of rival views seem to obey Bertrand Russell's law that subjective certainty is inversely proportional to objective certainty. Among those whose writings are gospel to some, anathema to others, and immensely stimulating to one and all, Dr. David Lack takes pride of place; and since he has again produced a book whose good qualities may dazzle the readers, the main function of the present review must be to cast a few shadows.[4]

The shadows cast by Chitty were so dark that they inspired at least one beginning graduate student in avian ecology to fervently pray that if he ever wrote a book, it would not be reviewed by Dennis Chitty. Chitty sent a reprint of the review to Lack with the handwritten inscription, "David: with all good wishes from your kindly critic. Dennis."[5] Butter wouldn't have melted in his mouth.

The balance of this chapter is devoted to the history of the controversy between Wynne-Edwards and Lack.

WYNNE-EDWARDS AND HIS GROUP SELECTION HYPOTHESIS

When *The Natural Regulation of Animal Numbers* was published in 1954, V. C. Wynne-Edwards was Regnus Professor of Natural History at the University of Aberdeen. As an undergraduate reading zoology at Oxford, he had been a member of the Oxford Ornithological Society when the society initiated the Oxford Bird Census, the precursor to both the Edward Grey Institute of Field Ornithology (EGI) and the British Trust for Ornithology (see Chapter 5). His reservations about Lack's views in *The Natural Regulation of Animal Numbers* actually predated its publication. Lack had presented a seminar at Aberdeen in 1953 expounding his theory that natural selection acts as a powerful agent to maximize the reproductive rate of the individual (see Chapter 3). Adam Watson, who was a student of Wynne-Edwards at the time, related that Wynne-Edwards leaned over to him during the seminar and said, "He obviously doesn't know anything about elephants or long-lived birds."[6] The battle was thus quietly joined.

Wynne-Edwards' first salvo was launched at the International Ornithological Congress (IOC) held in Basel, Switzerland, in 1954. He presented a paper entitled, "Low Reproductive Rates in Long-lived Birds, Especially Sea-birds," in which he asserted that in many long-lived birds, including penguins, albatrosses, and a number of large raptors, the reproductive rate is suppressed by several adaptations to prevent overpopulation. "Six such adaptations, all acting mutually to reinforce one another in depressing the recruitment rate,"[7] were then considered: (1) the minimum clutch size of 1 found in many seabirds, (2) failure to replace lost eggs, (3) prolongation of the period of parental care (i.e., long nestling periods), (4) reproductive cycles exceeding 1 year, (5) deferred maturity, and (6) sociable breeding habits (i.e., colonial breeding). In the discussion, Wynne-Edwards began to develop his more general ideas about population regulation:

It is theoretically possible to regulate numbers in the population by density-dependent control of recruitment alone... Control of this sort could be largely "intrinsic," that is, depending for its operation on the behaviour-responses of the members of the populations themselves. . . . Such birds as albatrosses, condors and the like do not in their adult stages attract specialized predators at all, and the advantage of having some autonomic control of numbers appears to be the more important.[8]

One salient example of "autonomic control" that Wynne-Edwards identified was territorial behavior in birds:

Bird-populations are generally prevented from exceeding the optimum density within their proper habitat by specially developed checks, the commonest of which is the holding of territory. Territory size is apparently correlated with the potential food supply and reflects the productivity of the habitat, with the result that the optimum density is not exceeded and the stock of food is safeguarded.[9]

This assertion would prove to be one of the major bones of contention in the ensuing debate between Wynne-Edwards and Lack.

Wynne-Edwards' first direct response to *The Natural Regulation of Animal Numbers* was his review of the book in *Discovery*, mentioned earlier (see also Chapter 5). Further elaboration of his own ideas came in subsequent presentations, one to the IOC in Helsinki in 1958 and the other to the Centennial Meeting of the British Ornithologists Union (BOU) held in London in 1959. In the first presentation, entitled, "The Over-fishing Principle Applied to Natural Populations and Their Food-resources: And a Theory of Natural Conservation," he used the overfishing of whales and some fish stocks by humans as an analogy for a proposed self-regulatory system in animal populations. He then stated, "It is possible to put forward a general theory to explain how overfishing is prevented in natural populations [that] may very readily be seen to accord in quite a striking manner with the known facts."[10] His proposal was that direct competition among individuals for food, which inevitably leads to overfishing, is replaced by a "purely conventional substitute" involving social interactions among the individuals of a population. Winners in these conventional competitive interactions are ensured of access to food, whereas losers are denied access to an adequate food supply and consequently fail to breed, emigrate, or die.

> Among all the higher animals such as birds competition for this right assumes two apparently distinct forms. . . . [T]he first form is identifiable as a contest for the use of real property, and the second as a contest for a sufficient status in the social hierarchy or peck-order.[11]

He identified territorial behavior in birds as a well-known example of the first form of conventional competition. Although he stated that natural selection would favor populations that were able to regulate their size so as not to overexploit their food resources, he was not specific in describing the mechanisms by which such characteristics could have evolved.

Wynne-Edwards was unable to attend the Centennial Meeting of the BOU because he was holding a visiting professorship at the University of Louisville in the United States at the time of the meeting. His paper, "The Control of Population-Density Through Social Behaviour: A Hypothesis," was therefore delivered by his colleague and former student, George Dunnet. The paper began by describing the results of a 1920s study by P. Jesperson in which the number of pelagic birds observed per day and the volumes of surface plankton (the base of the food chain for the birds) were sampled over a large part of the open ocean in the North Atlantic. Both the number of birds observed and the plankton volumes varied by more than 100-fold, but there was a highly significant positive correlation between the two. Wynne-Edwards attributed this pattern of dispersion in the birds to evolved social interactions that enabled them to match their numbers to the food availability in an area and thus prevent "overfishing" of the food resources. He coined a new term, "epidiactic demonstrations," for these interactions that serve to communicate among members of a population information about population density relative to local food availability. He did not identify the evolutionary mechanism by which such epidiactic demonstrations had developed but concluded the paper by saying in reference to his hypothesis, "A much fuller treatment of its implications is in active preparation."[12]

Indeed that was the case. *Animal Dispersion in Relation to Social Behaviour* ran to 653 pages when it was published 3 years after the BOU meeting. The opening chapter of the book was devoted to outlining Wynne-Edwards' theory of intrinsic population regulation. The chapter began, "Animal dispersion may be defined as comprising the placement of individuals and groups of individuals within the habitat they occupy, and the processes by which this is brought about."[13] Wynne-Edwards acknowledged the stimulus provided him by Lack's *The Natural Regulation of Animal Numbers* by noting that the last chapter of that book was entitled "Dispersion." The theory he outlined was a frontal assault on many of Lack's conclusions, however.

The chapter continued with extended descriptions of the Jesperson study demonstrating the relationship between the dispersion of pelagic seabirds and surface plankton volumes in the open ocean (the subject of the BOU presentation) and the overfishing analogy described in the Helsinki presentation. After identifying five "lessons" to be learned from the history of human overfishing, he stated:

> The importance of these inferences lies, of course, in the fact that there is no difference between man exploiting fish or whales and any other predator exploiting any other prey. All predators face the same aspects of exactly the same problem. It is impossible to escape the conclusion, therefore, that *something must, in fact, constantly restrain them, while in the midst of plenty, from over-exploiting their prey.* Somehow or other 'free enterprise' or unchecked competition for food must be successfully averted, otherwise 'overfishing' would be impossible to escape. . . . Such *prima facie* argument leads to the conclusion that it must be highly advantageous to survival, and thus strongly favoured by selection, for animal species (1) to control their own population-densities, and (2) to keep them as near as possible to the optimum level for each habitat they occupy.[14]

Wynne-Edwards then proposed another analogy, the concept of homeostasis from the science of physiology, to explain how "species" might achieve these two objectives. He referred to the self-regulating mechanism of populations as "population homeostasis" and stated, "Indeed this new manifestation of homeostasis in the process of life seems unlikely to remain long in doubt."[15] Physiological homeostasis requires feedback mechanisms by which the individual detects deviations from the homeostatic state, enabling it to make appropriate adjustments to return to the desired state. "In the balance we are considering here it is postulated that population-density is constantly adjusted to match the optimum level of exploitation of available food resources."[16]

Wynne-Edwards proposed that social behavior had evolved in animals as a means of providing the necessary feedback for them to achieve population homeostasis: "Putting the situation the other way round, a society can be defined for our purposes as *an organization capable of providing conventional competition.*"[17] He reiterated points made in his Helsinki presentation that in higher animals such as birds, this conventional competition is represented by social interactions such as territorial behavior and dominance hierarchies but also includes such behaviors as the dawn chorus in birds and communal roosting. All were capable in his view of providing feedback to the population about population density in relation to the available food resources, permitting the population to respond appropriately: "The actual regulation of population-density is largely a matter of exercising control over recruitment and loss in the population."[18] Wynne-Edwards now referred to the social behaviors that contribute the feedback for population homeostasis as epidiactic "displays" (rather than demonstrations).

Wynne-Edwards then proceeded to describe the linchpin of his theory: group selection. He recognized that selection at the individual level (traditional Darwinian selection) was incapable of explaining the evolution of traits that would permit individuals to forego their own well-being in favor of the well-being of the group:

Evolution at this level can be ascribed, therefore, to what is here termed group-selection—still an intraspecific process, and, for everything concerning population dynamics, much more important than selection at the individual level. The latter is concerned with the physiology and attainments of the individual as such, the former with the viability and survival of the stock or the race as a whole. When the two conflict, as they do when the short-term advantage of the individual undermines the safety of the race, group-selection is bound to win, because the race will suffer and decline, and be supplanted by another in which antisocial advancement of the individual is more rigidly inhibited.[19]

Group selection was not only the linchpin of Wynne-Edwards' theory of population homeostasis, but it would prove to be the principal focus of Lack's attack on his ideas.

The remaining 21 chapters of *Animal Dispersion in Relation to Social Behaviour* were devoted to describing applications of the theory to behaviors in a wide range of species, from the daily vertical movements of plankton to the mysterious aerial acrobatics of large, sometimes massive, flocks of European starlings before they enter their communal roosts. Wynne-Edwards identified both of these behaviors as epidiactic displays, providing feedback to the respective populations as a means of achieving population homeostasis. He described many other behaviors across the animal kingdom as examples of epidiactic displays. Other plausible functions had previously been identified for some of these behaviors, but some, Wynne-Edwards claimed, were inexplicable except in terms of his theory.

Some reviews of *Animal Dispersion in Relation to Social Behaviour* were quite favorable, such as those in the leading ornithological journals in both North America and Great Britain. Writing in *Auk*, Helmut Buechner chided Wynne-Edwards for including too much extraneous material but concluded: "Nevertheless, the book will provide a useful background of knowledge and orientation, as well as stimulation, for future research."[20] He did not mention group selection or its proposed role in the evolution of population homeostasis. Max Nicholson, writing in *Ibis*, did refer without comment to the role of group selection in Wynne-Edwards' proposed mechanism of population regulation, and he concluded his review by saying:

> Whether, like the present reviewer, the reader broadly agrees with the author's theory, or not, there can be no doubt that Professor Wynne-Edwards has made one of the most important of all contributions to the study of animal populations, and that all future discussion of the topic must be deeply influenced by this outstanding and richly illuminating book.[21]

However, Charles Elton, reviewing the book for *Nature*, was less enthusiastic. After describing the basic tenets of Wynne-Edwards' theory, including the role of group selection, he wrote:

> The theory is set forth with enthusiasm, often pontifically (if a bishop can wear blinkers), sometimes in a sort of Messianic exaltation which admits of no other important processes affecting population-levels. The language is usually lucid and the great array of facts that has been unearthed very interesting in itself; but the reasoning behind the language is extremely involved, and . . . rather woolly, and the new dogma incorporated in it safeguarded by many careful side-steps that make one rather uneasy. . . . Prof. Wynne-Edwards has not, so far as I know, carried out any field research on natural population-control, and he therefore tends to over-simplify the way in which animals live. . . . The enormous weakness of this enormous book, so full of fascinating information, so impregnated with one particular teleological bias, is that it gives no single case-history of group selection.[22]

For the second time in 20 years, Elton and Lack were in agreement (see Chapter 3).

Animal Dispersion in Relation to Social Behaviour was chosen by the Institute for Scientific Information as its "Citation Classic" for the week of June 23, 1980, reporting that it had been cited more than 600 times since 1962. In an accompanying statement, Wynne-Edwards stated that the idea for the self-regulation of animal populations had come to him while he was preparing his review of Lack's *The Natural Regulation of Animal Numbers* (see Chapter 5). He concluded his statement as follows:

> The theory still rings so true and explains so much that it seems almost bound to be correct, and of course prudential adaptations are not the only authentic phenomena that continue to perplex evolutionists.[23]

THE GROUP SELECTION CONTROVERSY

Lack's response to the publication of *Animal Dispersion in Relation to Social Behaviour* was swift, and his opposition to Wynne-Edwards' theory of population regulation was direct and unwavering. He invited Wynne-Edwards to present a special lecture at the annual Student Conference in Bird Biology at the EGI (see Chapter 5). The topic of the 17th Student Conference, held at St. Hugh's College in early January 1963, was Population Dynamics. Wynne-Edwards made his presentation on the morning of the final day of the conference, and after a coffee break four respondents led a lengthy discussion, each beginning with a short presentation: Niko Tinbergen on territory, Robert Carrick on social breeding in the Australian magpie, Mike Cullen on group selection, and Philip Ashmole on oceanic birds. The discussion was lively and lasted the entire afternoon. Lack's very positive description of the discussion (see Chapter 8) may have emanated in part from the fact that there was a virtual "murderer's row" of supporters of individual selection in attendance who were aligned with his own dim view of Wynne-Edwards' group selection theory.

The next exchange was initiated by a letter by Wynne-Edwards that was published in *Nature* in 1964. The letter was in response to an article in *Nature* by Chris Perrins, one of Lack's D.Phil. students, on survival in relation to clutch size in the swift. Wynne-Edwards reinterpreted some of the results reported in Perrins' paper to support his theory that birds reproduced at lower than the maximum rate possible. Lack responded with a letter to *Nature* in which his iciness of tone is only thinly disguised. He chided Wynne-Edwards for not distinguishing between the meanings of the terms "clutch size" and "brood size," providing definitions to underscore his point, and then proceeded to explain why he believed that swifts were in fact maximizing their reproductive output. Regarding Wynne-Edwards' alternative theory, Lack responded with a criticism frequently reiterated in *Population Studies of Birds*:

Wynne-Edwards wrote that "on my alternative hypothesis that inter-group selection has fixed the recruitment range around the optimum level for recruitment to the population as a whole, the facts . . . seem reasonably intelligible." But he provided no positive evidence for this statement, and in particular he has given no supporting figures for correlated variations in fecundity recruitment or size of population. Hence there is no means of knowing whether recruitment is at the 'optimum level'.[24]

In a rejoinder published simultaneously with Lack's letter, Wynne-Edwards admitted that there was some confusion in his earlier letter but responded in kind to Lack regarding his attack on group selection:

> But notwithstanding the errors in the figures . . . Lack confuses three suggestions as to how this situation could be accounted for on the basis of natural selection operating at the level of the individual.[25]

The battle lines between the two were thus clearly demarcated.

In *Population Studies of Birds*, Lack's attack on Wynne-Edwards' theory was two-pronged: (1) an attack on Wynne-Edwards' assertion that many species, including long-lived birds, reproduce at lower than the maximum rate possible, and (2) an attack on the theory of group selection that Wynne-Edwards proposed to account for the evolution of reproductive rates below the maximal rate. His principal arguments were that Wynne-Edwards failed to provide any positive evidence for many of his assertions and that he repeatedly failed to mention plausible alternative hypotheses that had been proposed to explain many of the behaviors that Wynne-Edwards identified as epidiactic displays. His attacks on Wynne-Edwards' theory of population regulation were direct and often very pointed. Lack used terms such as "inconceivable," "highly improbable," and "ignorance" in his critique. He summarized his arguments against Wynne-Edwards as follows:

> To summarize, there is indeed a phenomenon of dispersal, whereby birds and other animals modify their population density in relation to the food supply through movements and in other ways, particularly while breeding, but this phenomenon is not nearly as extensive as Wynne-Edwards claimed. Also at least most of the behaviour considered epidiactic by Wynne-Edwards can be satisfactorily interpreted in other ways, and no positive evidence has been presented for an epidiactic function. Further there is no reason to suppose that reproductive rates have been evolved to balance mortality rates; reproductive rates are explicable through natural selection, and the balance

between birth-rates and dearth-rates can be attributed to density-dependent mortality. Finally, not only low reproductive rates, but also the types of behaviour discussed by Wynne-Edwards, including those which genuinely involve dispersion, are explicable through natural selection and there is no need to invoke group-selection.[26]

RAPPROCHEMENT IN THE SCOTTISH HIGHLANDS

Lack's pointed criticisms of the ideas put forward by Wynne-Edwards in his *magnum opus* resulted in a definite chilliness in their relationship. One of Wynne-Edwards' students commented that Wynne-Edwards tended to react defensively to criticism of his ideas, and the piercing attacks by Lack strained the relationship between the two men, who had long been friends. In the summer of 1968, Lack decided to try to mend the frayed connection. He contacted Adam Watson about the possibility of arranging for a joint outing in Scotland between the Lack and Wynne-Edwards families during the summer holiday. Watson was happy to comply with David's request, and he organized a day outing in the Cairngorms, in the eastern highlands of Scotland.

The excursion took place in July 1968, and Watson drove the Land Rover carrying David, Elizabeth, and their four children on their approach to the River Dee side of Beinn a' Bhuird, a mountain in the eastern part of the Cairngorms. They met Wynne-Edwards at a prearranged location to hike together on the mountain. The ostensible objective of the hike was to look for the rare highland or brook saxifrage. Andrew Lack recalled meeting Wynne-Edwards for the first time: "I remember an extraordinarily fit, wiry, active and most unassuming man, with a gait that looked as if he could walk all day (he could!)."[27] The morning search proved unsuccessful. Adam Watson described the ensuing events:

At lunch we sat by a mossy spring admiring the view. David tentatively mentioned population regulation, and Wynne tentatively replied. The discussion then ended as David spotted [a] hen dotterel beside us. He certainly was showing his prowess as an observer. We then spread out for another plant search, walking slowly, but VCWE [V. C. Wynne-Edwards] announced that he had not had enough exercise and would run down the hill and see us in the glen. The Lacks and I continued searching, and eventually found the brook saxifrage in flower on a cold cliff. Elation! Later, when we picked up Wynne, hot and sticky after running in the sun, a beaming David told him we had found the plant! It was a moment of supreme one-upmanship, which had its funny side to VCWE too, after his initial disappointment. More importantly, the ice was now well and truly broken.[28]

Their differences regarding population regulation and group selection remained unresolved, but the personal relationship between Lack and Wynne-Edwards was "well and truly" mended.

Monogamy, in birds and men, tends to be dressed in drab colours, but it is far more frequent than the exotic alternatives.... [O]ver nine-tenths of all birds are monogamous, in nidicolous species well over nine-tenths, and even in nidifugous species, many of which do not feed their young, nearly four-fifths.... The above proportions are unlikely to be changed by further research.

—*Ecological Adaptations for Breeding in Birds*, p 148

10

Ecological Adaptations for Breeding in Birds

⌒⎯⎯⎯⎯⎯⎯⎯⎯⎯⎯⎯⎯⎯⎯⎯⎯⎯⎯⎯⎯⎯⎯⎯⎯⎯

IN THE LATE 1950s, a doctoral student working under the supervision of William Thorpe and Robert Hinde at Cambridge began pioneering studies on the ecological correlates of various breeding behaviors, such as nesting dispersion and type of pair bond, in the weaverbirds of Africa. His name was John H. Crook, and the results of his studies were to inspire David Lack to undertake a comprehensive study of the ecological correlates of breeding behaviors in birds. The results were published in Lack's second most frequently cited work, *Ecological Adaptations for Breeding in Birds*, which was published in 1968 by Methuen. David acknowledged his debt to John Crook by sometimes referring to it as "Crook's book."

Lack chose seven breeding characteristics for his comparative analysis: nesting dispersion, pair bond, clutch size, egg size, incubation period, nestling period, and age at first breeding. He also separated the birds into six ecological groups for analysis. The first and largest of these groups were the passerines (songbirds), primarily terrestrial birds that have nidicolous young (meaning that hatchlings are helpless, are confined to the nest, and must be fed and brooded by one or both of the adults). The second and next largest group comprised other nidicolous land birds, such as woodpeckers and doves. Together, these two groups made up more than 85 percent of the approximately 8500 species of birds recognized at the time. The other four groups were ground-living land birds with precocial (nidifugous) young, wading and littoral water birds, freshwater aquatic birds, and marine aquatic birds. He also chose the subfamily (or family, for monotypic families) as the taxonomic unit for his comparisons.

The text of *Ecological Adaptations for Breeding in Birds* was divided into two parts. The first part dealt with patterns of nesting dispersion and the pair bond in the different ecological groups. Chapters were devoted to each group, and there were several additional chapters dealing with special groups including the weaverbirds, brood parasites, and cooperatively breeding species. Part One concluded with two chapters summarizing the major conclusions regarding nesting dispersion and the pair bond in birds. Part Two dealt with clutch size, egg size, and growth rates (incubation period, nestling period, and age at first breeding). Again chapters were devoted to each ecological group, with additional chapters on variations within some seabird families, growth rates in Procellariiformes (albatrosses and shearwaters), special problems of eggs, and timing of breeding. Part Two ended with a concluding chapter in which Lack summarized the major findings of his analysis:

> In general, as will have become clear, I consider that all the breeding habits and other features discussed in this book have been evolved through natural selection so that, in the natural habitats where they were evolved, the birds produce, on the average, the greatest possible number of surviving young. The evidence supporting this view is much stronger for certain of these features than others, and as yet is based mainly on comparisons between species and so is circumstantial, but the general picture seems coherent and convincing. The main environmental factors concerned in this evolution are the availability of food, especially for the young and to a lesser extent for the laying female, and the risk of predation on eggs, young and parents. The effects are in part counteracting, so that each adaptation is a compromise between conflicting advantages, and further, each adaptation influences many others, so that the end-result is due to a complex of interacting factors.[1]

This conclusion demonstrated that in some respects Lack's views had changed little from those expressed in his classic clutch size papers of 1947–1948 (see Chapter 3), in which he asserted that natural selection would always operate to maximize the reproductive output of the individual and that food was usually the limiting factor determining how many young could be successfully reared. However, it also showed that he now recognized that clutch size was only one of a suite of reproductive attributes in the reproductive strategy of a species. This expansion of Lack's clutch size hypothesis might be expressed as follows: The reproductive strategy of a species has been evolved by natural selection to maximize the reproductive potential of the individual.

The conclusion expressed in the epigraph to this chapter, that more than 90 percent of bird species are monogamous, was soon to be challenged by the sociobiological revolution of the early 1970s. Ironically, this revolution emerged in part from the implications of the idea that natural selection operates solely at the level of the individual—an idea consistently promulgated by Lack himself, not only in his clutch size hypothesis but also prominently in *Population Studies of Birds* (see Chapter 9). Early theorists in the movement, such as Robert Trivers, began to explore the possibility that competing selection pressures were operating on offspring and their parents ("parent-offspring conflict"), as well as between the members of a monogamous pair, leading to mixed-mating strategies. Definitive support for the latter idea had to await the application of DNA-fingerprinting technology to parent birds and their putative offspring in the mid-1980s. The technique, first applied to house sparrows, showed that in many monogamous species, 10 to 15 percent or more of the offspring were not sired by the putative father, indicating that successful extra-pair copulations were frequently occurring. The results led to the use of the term "social monogamy" to describe the relationship between members of a supposedly monogamous pair.

Lack frequently expressed skepticism about the results of laboratory or caged studies of birds, believing that birds often behaved in abnormal ways under such artificial conditions. Had he been less skeptical of such studies, he might well have discovered this mixed-mating strategy of supposedly monogamous males much earlier, because he had once observed a male robin in an outdoor aviary at Dartington Hall copulating with a female other than its mate.[2]

Ecological Adaptations for Breeding in Birds received numerous favorable reviews. Bertram G. Murray, Jr., writing in the *Auk*, concluded his review by saying, "This book must be read and reread, studied and restudied by anyone planning to do serious research in this field. If you do not already have a copy, order one now!"[3]

LACK'S D.PHIL. STUDENTS

Academic lineages trace the relationships of doctoral students and their supervisors the students' own doctoral students, and they sometimes yield interesting insights into the development and perpetuation of scientific ideas. David Lack supervised 19 D.Phil. students, and an examination of their professional contributions and those of their doctoral students suggests that Lack's "intellectual genes" have been widely disseminated and have had an enormous influence on developments, not only in ornithology and evolutionary ecology but in many related fields, particularly behavior and conservation biology.

As a supervisor, Lack usually permitted students to pursue their own interests, particularly if they arrived with a strong interest in a particular bird species. He sometimes had projects that he encouraged incoming students to take on, but if their true interests lay elsewhere, he would release them to pursue those interests. In general, he had a "hands off" approach, allowing students to develop their own projects. One said that Lack urged his students to always "define their question" and stated that the questions could not be found by "simply walking through the woods." Another student said that Lack helped his doctoral students "sort out their ideas" and encouraged them to accept the "simplest explanation, which was probably best." Several of his students remarked that he was very helpful in commenting on drafts of their theses, although one reported that Lack said he would read the thesis only once, in draft or in its final form (the student chose the latter option).

In the following accounts of Lack's D.Phil. students, the dates associated with each name are those from their D.Phil. theses and may not correspond to the year of actual conferral of the degree.

ROBERT A. HINDE (1951)

David Lack's first D.Phil. student arrived in the Edward Grey Institute of Field Ornithology (EGI) in 1948 after completing his undergraduate degree at St. John's College, Cambridge. He had attended the first Student Conference in Bird Biology in December 1946 while still an undergraduate. During World War II, he had served in the R.A.F. as a PBY airplane pilot flying coastal watch missions. Lack had hoped that he would do a comparative food study of jackdaws and rooks, but Robert's interests were more behavioral than ecological, and he chose to work instead on the Wytham Woods tits. Although Lack was his supervisor, Robert worked closely with Niko Tinbergen, who relocated to Oxford from the Netherlands in September 1949. His thesis was entitled, *A Comparative Behaviour Study of the Paridae*.

After Hinde completed his doctorate, W. H. Thorpe lured him back to Cambridge to serve as curator of the newly established Ornithological Field Station on the Madingley estate 6 miles outside Cambridge. The initial research program of the station centered on the study of bird song, but the scope of research quickly broadened to encompass imprinting, avian neuroendocrinology, sensitive periods during development, primate behavior, and social development in children. In recognition of these changes, the university renamed the station the Sub-Department of Animal Behaviour in 1960. Students associated with the station included Peter Marler, known for his pioneering work on bird song, and Jane Goodall and Diane Fossey, who began their well-known field studies on chimpanzees and gorillas, respectively, while at Madingley.

The broadening research program at Madingley also contributed to the establishment by the Medical Research Council of the MRC Unit on the Development and Integration of Behaviour, which Hinde served as Director from 1970 until 1989. His own interests had broadened in tandem with those at Madingley, and he became increasingly interested in human social behavior. He also became an outspoken war critic, as evidenced by the title of a book he co-authored with Joseph Rotblat, *War No More: Eliminating Conflict in the Nuclear Age* (Pluto, 2003). Hinde authored several other books, including *Biological Bases of Human Social Behaviour* (McGraw-Hill, 1975), *Ethology* (Fontana, 1986), *Individuals, Relationships and Culture: Links between Ethology and the Social Sciences* (Cambridge University Press, 1988), *Relationships: A Dialectical Perspective* (Psychology Press, 1997), *Why Gods Persist: A Scientific Approach* (Routledge, 1999), *Why Good is Good: The Sources of Morality* (Routledge, 2002), and *Bending the Rules: Morality in the Modern World—From Relationships to Politics and War* (Oxford University Press, 2007). He also wrote a leading textbook on animal behavior and edited numerous other volumes.

Hinde was elected a Fellow of the Royal Society in 1974. He was also elected an Honorary Fellow of the British Academy, Honorary Foreign Associate of the National Academy of Sciences (USA), and Foreign Honorary Member of the American Academy of Arts and Sciences. He served as Master of St. John's College, Cambridge, from 1989 to 1994.

In 2012, Hinde is retired and living in Cambridge; he remains active in research and writing.

MONICA M. BETTS (1952)

Monica Betts was one of David's first D.Phil. students, and she was certainly one of the first women to earn a doctorate in ornithology. She came to the EGI in 1948 as Peter Hartley's field assistant after completing a B.Sci. degree at University College, London. She soon obtained a 5-year grant from the Agriculture Research Council to work on the tit project in Wytham Woods, and she completed a thesis entitled, *The Availability of Food and Predation by the Genus Parus*.

After completing her doctorate, Betts received a second 5-year grant to work on tits in the Forest of Dean. She also worked with John Gibb on the Breckland Research Project funded by the Nature Conservancy (see Chapter 5). She ended her teaching career at Oxford Brookes University and is still living in Oxford.

DAVID W. SNOW (1953)

Robert Hinde was not the only one of Lack's D.Phil. students to have attended the first Student Conference in December 1946. David Snow was a newly arrived undergraduate at New College, Oxford, whose education, like Hinde's, had been interrupted by World War II. Following the study of classics at Eton, he had been awarded a scholarship in classics at New College, but the Royal Navy preempted his matriculation. He arrived in Oxford in October 1946 having served aboard a destroyer in both the North Atlantic and the Far East for 3 years. In that interim, he had also decided to read zoology instead of classics. Because New College had no tutor in zoology at the time, his tutelage by David Lack and his involvement with the EGI began almost immediately.

Snow joined the Institute as a research student in 1949, after receiving the University's Christopher Welch Prize on completion of his baccalaureate. His doctoral research, on the ecology and systematics of the Old World tits, culminated in his thesis, *Systematics and Comparative Ecology of the Genus* Parus *in the Palearctic Region*. After completion of his doctorate, Snow remained at the EGI as a senior research officer until 1957, with his principal research work focusing on blackbirds in the Botanic Garden. This research culminated in the publication of *A Study of Blackbirds* (Allen and Unwin, 1958).

Snow left the EGI in February 1957 to become director of the New York Zoological Society's Field Station in Trinidad. His fiancée, Barbara Whitaker, who was also an ornithologist, soon joined him there, and they married in Trinidad. They began a long collaboration of the study of frugivorous birds and fruit-bearing plants that ultimately led to the jointly authored *Birds and Berries: A Study of Ecological Interaction* (T. & A.D. Poyser, 1988).

Snow returned to the EGI as a senior visitor in 1962 but left in January 1963 to become director of the Charles Darwin Research Station in the Galapagos. In 1964, he became research director of the British Trust for Ornithology (BTO), a post he held until 1968, when he became director of bird collections at the British Museum of Natural History. He remained in that position until his retirement in 1984.

Snow wrote *The Web of Adaptation: Bird Studies in the American Tropics* (Quadrangle, 1976) and *The Cotingas* (Cornell University Press, 1982). He was also an outstanding editor, serving as editor of *Ibis* from 1968 to 1973 and as lead editor of *The Birds of the Western Palearctic* (Oxford University Press, 1997). He also served as president of the British Ornithologists' Union (BOU) from 1987 to 1991. He received the Godman-Salvin Medal of the BOU in 1982, and he and Barbara shared the William Brewster Medal of the American Ornithologists' Union in 1972. David Snow died in February 2009.

JOHN A. GIBB (1953)

After serving for 4 years in Malta with the Royal Artillery during World War II, John Gibb returned to Oxford as a field assistant at the EGI in July 1946. During the next few years, along with Monica Betts and Robert Hinde, he did much of the field work in the project on tits begun in 1947 in Wytham Woods. His thesis, *Factors Governing Population Density of Birds of the Genus* Parus, signaled his lifelong interest in factors regulating population numbers. After completion of his doctorate, he continued working in the EGI for 4 years as senior research officer. He won the Bernard Tucker Medal of the BTO in 1956 in recognition of his outstanding contributions to British ornithology.

In 1957, John migrated to New Zealand to study rabbit populations for the agency that was to become the Ecology Division of the Department of Scientific and Industrial Research. He became director of the division in 1965, a post he held until his retirement in 1981. His own work there centered on rabbit control and predator-prey interactions, but he also helped design projects on introduced bird pests and on the conservation of native bird species. He died in New Zealand in 2004 at the age of 85.

WILLIAM J. L. SLADEN (1954)

Bill Sladen was a somewhat unusual D.Phil. student, although by no means the only such to be accepted by Lack into the EGI. He had completed an M.D. degree in London in the 1940s and had accompanied Sir Vivian Fuchs on his first Antarctic expedition as medical officer. The expedition, which departed England in December 1947, lasted for 2 years. Sladen's interest in penguins developed during this expedition, and this interest brought him to the EGI to write up his penguin research. He chose the EGI in part because of Lack's reputation but also because he had heard that the library there was the best open library available, and because the well-known New Zealand penguin biologist, L. E. Richdale, was Nuffield Research Fellow in the EGI at the time. Sladen was a practicing plastic surgeon at the time he arrived in Oxford.

Although Sladen remembered Lack as a "fantastic editor," he also described him as possessing many of the stuffy characteristics of the typical Oxford don. He found that Lack insisted on maintaining the strict hierarchical relationship between professor and student and tended to dominate the conversations at EGI tea times, usually discussing topics that were related to whatever he was working on at the time. Sladen believed that David found it somewhat offensive that he, a student, had a finer car (a Ford Prefect) than David's own Morris Minor.

Sladen's thesis was entitled, *The Biology of the Pygoscelid Penguins*. Among many other significant insights into penguin biology, he was the first to show that parent penguins recognize their own chicks in the seemingly chaotic crèches comprising hundreds of chicks. When he left the EGI for a Rockefeller Fellowship in the United

States, where he became involved in the early stages of the U.S. Antarctic Research Program, Sladen took more than a freshly minted Oxford degree. Despite the well-known locked door that separated Lack's EGI from Elton's Bureau of Animal Population (BAP) in their adjacent quarters at the Botanic Garden, there was much informal contact among the students and staff of the two institutions. Bill met and courted Elton's field assistant, Brenda Macpherson, while at the EGI, and she became his wife.

Sladen subsequently received an appointment to the faculty of the School of Medicine at Johns Hopkins University in Baltimore, Maryland. He continued his research in the Antarctic, banding some 50,000 penguins and 60,000 albatrosses in the course of long-term studies there. He also initiated extensive research in the Arctic, working primarily on geese and swans. He had first visited Greenland during his time at the EGI, accompanying an expedition led by Peter Scott to band pink-footed geese. He retired from Johns Hopkins as Emeritus Professor of Microbiology and Immunology.

After retirement, Bill concentrated on his work with swan conservation, particularly that of the North American trumpeter swan. He founded the Airlie Environmental Studies Program at the Airlie Institute in Virginia and served as its director from 1989 to 2006. One of the projects that he oversaw served as the inspiration for a popular motion picture, *Fly Away Home* (1996). That movie chronicled the training of young Canada geese to imprint on an ultralight aircraft, which then led them on a several-hundred-mile migration from Canada to the southeastern United States. The project was actually being performed by conservation biologists to test the hypothesis that young waterfowl such as geese and swans, which learn a traditional migratory flight route by following their parents, could be taught such a route using the ultralight as a surrogate parent. As anyone who has seen *Fly Away Home* can tell you, the experiment proved successful, and the majority of the young geese returned to Canada on their own the following spring. However, most cannot tell you "the rest of the story": Because geese imprinted on humans were unwanted as breeding birds in Canada, the returnees became martyrs for a greater conservation cause. They were shot, and their bodies now lie in state in museum trays in the Royal Ontario Museum, Toronto. Since then, the technique of imprinting young waterfowl on ultralight aircraft has been successfully employed to teach both young trumpeter swans and whooping cranes new migratory routes in efforts to establish new breeding populations of these two endangered species.

JAMES D. LOCKIE (1954)

James Lockie arrived at the EGI as a research student in October 1951 after obtaining a B.Sci. degree. After completing his D.Phil. thesis, *The Feeding Ecology of the Jackdaw, Rook, and Related Corvidae*, he left the institute in August 1954 to accept a position with the Nature Conservancy in Edinburgh, where he remained throughout his career. He rose to the position of Chief Scientific Officer there and was also a lecturer in Animal Ecology and Wildlife Management at Edinburgh University.

Much of his research focus was on avian and mammalian predators, including eagles, owls, weasels, martens, and foxes. During the 1960s, he worked on the effects of organochlorine residues on breeding success in golden eagles, and his was one of the pioneering studies that ultimately led to the banning of DDT and other organochlorine-based pesticides. Lockie's work with Nature Conservancy also led him into problems of rural land use in Scotland. He co-authored (with Donald McVean) the book, *Ecology and Land Use in Upland Scotland* (Edinburgh University Press, 1969).

BERNARD STONEHOUSE (1957)

Born near Hull, Yorkshire, in 1926, Bernard Stonehouse had had a full life's range of experiences before he arrived at the EGI as a research student in 1955. Like Bill Sladen before him, Stonehouse's route to the EGI was circuitous and included two stints in the Antarctic. He was a fourth form student at the beginning of World War II and, like many other British children, was evacuated from Hull during the German bombing of major ports and industrial centers. He subsequently joined the Air Training Corps and was trained as a pilot. He briefly attended University College, Hull, reading zoology and botany, but then joined the Falkland Island Dependencies Survey in 1946 and was assigned to Base E, the large sledding station on Stonington Island in the Antarctic. While serving as copilot of the station's airplane, he survived a crash landing far from the station and 7 days of trudging through slushy snow with limited provisions before being rescued. The summer he was scheduled to be relieved of duty on the island, the ship bringing supplies and replacement personnel was unable to break through the ice, and he was forced to remain through a third Antarctic winter. (He and two of his co-workers thus became the first people to spend three consecutive winters south of the Antarctic Circle). He spent that third winter studying a breeding colony of emperor penguins, work that would ultimately lead him to the EGI. After leaving the service, he began undergraduate studies at University College, London, earning a B.Sci. degree in zoology in 1953. While there, he attended a Student Conference at the EGI and read a paper describing his penguin research. David Lack encouraged him to publish his research and invited him to return to the EGI as a doctoral student after he had completed his degree.

Stonehouse received a grant for a 2-year study of king penguins on South Georgia following the completion of his undergraduate education, and he met with David to

discuss his project before departing for the island in September 1953. He returned to
the EGI in August 1955 to write his thesis, *Breeding Behaviour in the Genus* Aptenodytes, *with Particular Reference to* Aptenodytes patagonica.

After completion of his doctorate, Stonehouse received a half-time appointment
as a research officer in the EGI and was assigned to head the Ascension Island expedition (see Chapter 8). The 2-year expedition to study the seabirds of the remote
equatorial island was sponsored by the BOU as part of its centenary celebration.
Stonehouse was accompanied by his wife, Sally (who had been assistant librarian in
the EGI), and two EGI research students, Philip Ashmole and Doug Dorward.

After returning from Ascension Island, Stonehouse remained in his post as
research officer in the EGI until August 1960, when he left to become a senior lecturer in zoology at the University of Canterbury in Christchurch, New Zealand.
During his 8 years in that country, he developed two major research programs, one
involving studies of several introduced mammals in New Zealand and another
studying penguins, seals, and human impacts in McMurdo Sound, Antarctica.

In 1968, Stonehouse left New Zealand for a 6-month stint at the Scott Polar
Research Institute in Cambridge and a semester-long visiting professorship at Yale
University. He then accepted a 1-year appointment to a Canadian Commonwealth
Research Fellowship at the University of British Columbia, studying Dall sheep in the
Yukon. He returned to Britain in 1970 and taught biology for a short time at a public
school in Scotland before accepting an offer from the recently established and innova-
tive University of Bradford to develop its School of Studies in Environmental Science.
The resultant 4-year undergraduate degree program was cross-disciplinary and provided students with strong grounding in ecology, human biology, social history, environmental law, economics, chemistry, and sociology; it proved to be highly successful.

Bernard left Bradford in January 1983 and returned to the Polar Research Institute
as editor of its journal, *Polar Record*, a post he held until his retirement in 1992.
During his long academic career, Stonehouse had discovered that he had the ability
to write with an easy style that enabled him to communicate complex scientific concepts to a general audience. No doubt inspired by David Lack's example, he wrote
(or co-authored) dozens of books for the general public as well as several technical
books. Among the latter are *The Biology of Penguins* (Macmillan, 1975), *Evolutionary
Ecology* with Chris Perrins (Macmillan, 1977), *The Biology of Marsupials* with Desmond Gillmore (Macmillan, 1977), *Animal Marking: Recognition Marking of Animal Research* (Macmillan, 1978), *Biological Husbandry: A Scientific Approach to
Organic Farming* (Buttersworth-Heinemann, 1981) and *North Pole, South Pole: A
Guide to the Ecology and Resources of the Arctic and Antarctic* (McGraw-Hill, 1990).
The books for a more general audience included *Animals of the Arctic: The Ecology of
the Far North* (Henry Holt, 1971), *Animals of the Antarctic* (Holt, Rinehart and

Winston, 1972), and *The Last Continent: Discovering Antarctica* (SCP Books, 2000), as well as textbooks such a *The Way Your Body Works* (Crown, 1974) and *Polar Ecology (Tertiary Level Biology)* (Routledge, 1989). He continued to write prolifically in retirement.

For services to the BOU, Stonehouse was awarded the Union Medal in 1971.

N. PHILIP ASHMOLE (1961)

Philip Ashmole was appointed as a research student at the EGI in September 1957, after completing his B.Sci. in zoology at Brasenose College, Oxford. Although he was initially slated to work on jackdaws, Bernard Stonehouse selected him for the Ascension Island expedition, and beginning in September 1957 he spent almost 2 years there along with Stonehouse, his wife Sally Stonehouse, and Doug Dorward.

Ashmole's research on Ascension Island focused on the annual cycle and molt in terns, and when he returned to Oxford in 1959, he wrote his D.Phil. thesis, *The Biology of Certain Terns: With Special Reference to the Black Noddy* Anous tenuirostris *and the Wideawake* Sterna fuscata *on Ascension Island*. He also married Myrtle Goodacre, whom he had met at a Student Conference while both were still zoology undergraduates (she at University College, London). She had been serving as assistant librarian in the Alexander Library of Ornithology during Philip's sojourn on Ascension Island.

After completing his doctorate, Ashmole was appointed a university demonstrator, and he also continued as a research officer in the EGI. David Lack, working with G. Evelyn Hutchinson, helped arrange a summer research fellowship at Yale University, which eventually resulted in Ashmole's spending a year at the Bishop Museum in Hawaii studying the effects of nuclear test blasts on terns and other birds of Christmas Island. After that year, Ashmole was appointed Assistant Professor of Biology at Yale, a post he held until he accepted a similar position at Edinburgh University in 1972. He remained in the latter post until his retirement in 1992.

At the end of the Ascension Island expedition, Ashmole and Dorward had spent a month on St. Helena examining the fossil birds of the island. Philip and Myrtle resumed research there after he took the post at Edinburgh, focusing on both fossil birds and subterranean insects. They co-authored *St. Helena and Ascension Island: A Natural History* (Anthony Nelson) in 2000. The couple also became increasingly interested in conservation issues, which ultimately led to their involvement in a native woodland restoration project in the southern uplands of Scotland, Carrifran Wildwood.

DOUGLAS F. DORWARD (1961)

Doug Dorward joined the EGI as a research student in October 1955, supported by a grant from Nature Conservancy. He had completed a B.Sci. degree from Edinburgh University, and he began research on jackdaws (following on the studies of J. D. Lockie). In 1957, he was selected to participate in the BOU Centenary Expedition to Ascension Island and joined other EGI personnel, B. Stonehouse, S. Stonehouse, and N. P. Ashmole, for the 2-year stint. There, he worked on the breeding biology and molt of two species of boobies, and he subsequently completed a thesis entitled, *Comparative Breeding Biology of Some Seabirds of Ascension Island with Special Reference to Two Species of* Sula *and the Fairy Tern.*

Dorward emigrated to Australia in 1962, taking a lectureship in the Department of Zoology at Monash University, where Jock Marshall was Head of the department. He worked on albatrosses and on the conservation of the endangered Cape Baron goose. He also did programs on conservation and ecology for Australian television and wrote the book, *Wild Australia: A View of Birds and Men* (Collins, 1977). He attained the rank of Associate Professor in Zoology at Monash before dying in 1981 at the age of 48.

CHRISTOPHER MYLES PERRINS (1963)

Chris Perrins arrived at the EGI as field assistant in September 1957, after reading zoology at Queen's College, London, and he has been a fixture there ever since. Fifty-six years later, you are liable to see him at tea time on almost any day that you visit the EGI. If you don't know him, he's the "large man, whose rear shirt-tail is nearly always protruding over his trousers," as he was described in his 1988 Godman-Salvin Medal citation.

Chris's primary duties as field assistant involved monitoring the population studies of tits in Wytham Woods and conducting the observations of common swifts in the University Museum tower. He soon expanded the tit studies considerably by adding many new nest boxes in another part of Wytham Woods (Great Wood), as well as in pine woodlands and urban areas. He also initiated brood-size manipulations of both tits and swifts (Figure 16). He became a research officer of the institute in 1960 and completed his D.Phil. 3 years later, with a thesis entitled, *Some Factors Influencing Brood-size and Populations in Tits.*

Chris remained at the EGI as a research officer, and was appointed senior research officer in 1968. He continued to study the tits of Wytham Woods but also began work on Manx shearwaters on Skokholm Island and mute swans in Dorset (in

FIGURE 16: David (left) and Chris Perrins looking down a chimney for chimney swift nests at Ithaca, New York, during the 1962 International Ornithological Congress. (Photograph courtesy of Lack family.)

collaboration with M. Ogilvie of the Wildfowl Trust). The latter studies eventually led to his innovative use of the centuries-old tradition of the annual swan-upping on the River Thames between London and Oxford to study swan populations there. These studies resulted in his appointment as Her Majesty's Swan Warden in 1983 and the awarding of the LVO (Lieutenant of the Royal Victorian Order) in 1987.

The year after David Lack's death in 1973, Perrins was appointed as the third Director of the EGI, a position he held until his retirement in 2002. By the time of his retirement, he had supervised more than 90 doctoral students in ornithology, no doubt producing more ornithological doctorates than any other supervisor.

Perrins authored (or co-authored) several books, including *Evolutionary Ecology* with Bernard Stonehouse (Macmillan, 1977), *British Tits* (Collins, 1979), and *Avian Ecology* (Chapman & Hall, 1983). He also edited numerous books, most prominently serving a co-editor of the nine-volume *Handbook of the Birds of Europe, the Middle East, and North Africa: The Birds of the Western Palearctic* (Oxford University Press).

Chris served as acting head of the Department of Zoology in 1991–1993, while Sir Richard Southwood was serving as Vice-Chancellor of Oxford University, and he was named Professor of Zoology in 1993. He received the Godman-Salvin Medal from the BOU in 1988, and he was elected a Fellow of the Royal Society in 1997.

IAN NEWTON (1964)

Ian Newton came to the EGI as a research student in 1961 after reading zoology at Bristol University. His first visit to the EGI had been to attend the 1959 Student Conference, at which the topic was Tropical Ecology. His D.Phil. thesis was entitled, *The Ecology and Moult of the Bullfinch*. After completing his doctorate, he remained at the EGI for 3 years as a research officer, continuing his work on bullfinches under sponsorship of the Agriculture Research Council.

Newton left the EGI in 1967 to work for the Nature Conservancy studying crop damage by geese in Scotland. After spending the 1970–1971 year in Canada studying ducks with the Canadian Wildlife Service, he joined the newly formed Institute of Terrestrial Ecology, an offshoot of Nature Conservancy, to study raptor populations. He was appointed head of the Subdivision of Animal Function in 1978, and he stayed with the institute until retiring in 2000. In retirement, he has remained active in research and writing.

Ian is the author of numerous books, including *Finches* (Collins, 1972), *Population Ecology of Raptors* (Poyser, 1979), *The Sparrowhawk* (Poyser, 1986), *Population Limitation in Birds* (Academic Press, 1998), *The Speciation and Biogeography of Birds* (Academic Press, 2003), *The Migration Ecology of Birds* (Academic Press, 2007), and *Bird Migration* (Collins, 2010). He has served as president of both the BOU and the British Ecological Society.

Ian was awarded honorary doctoral degrees from Oxford in 1982 and from Sheffield in 2001; he received the Union Medal of the BOU in 1987 and that group's Godman-Salvin Medal in 2010. The British Ecological Society awarded him its Gold Medal in 1989, and the American Ornithologist's Union presented him with its Elliott Coues Award in 1985. He was elected a Fellow of the Royal Society in 1993 and received an OBE (Order of the British Empire) in 1999.

DAVID C. SEEL (1965)

David Seel joined the EGI as a research student in October 1960, after completing a B.Sci. in zoology at Queen Mary College, London. His research on the breeding biology and molt of house sparrows and tree sparrows in Wytham Woods was sponsored by a Nature Conservancy fellowship. He was forced to interrupt his work for a year due to illness but resumed his research in April 1963 and subsequently completed his thesis, *The Breeding Ecology of the House Sparrow* (Passer domesticus).

Seel left Oxford in January 1965 to accept an appointment as a demonstrator in Zoology at the University College of Wales, Aberystwyth (now Aberystwyth University). He subsequently took a position in the Zoology Department at the University of Southampton for several years before moving to the Bangor Research Station of the Institute of Terrestrial Ecology. His principal research was on European cuckoos and on fluoride concentrations in birds of prey. He remained at the Bangor Station until the late 1980s but moved to the computing laboratory at University College of North Wales (now Bangor University) before his death.

PETER RICHARD EVANS (1965)

Peter Evans came to the EGI after completing B.A. and Ph.D. degrees in chemistry at Cambridge. While at Cambridge, he participated in pioneering studies on visible migration at Gibraltar with Ian Nisbet, work that contributed to his decision to pursue ornithology as a career. A Nuffield Foundation Biological Scholarship, given to applicants who wish to change fields, enabled him to pursue a D.Phil. at the EGI. His thesis was entitled, *Some Aspects of Bird Migration in Northeast England.*

Evans remained in the EGI as a research officer for 3 years after the completion of his doctorate, working primarily on pied flycatchers in the Forest of Dean and on the fat and protein reserves of yellowhammers in Wytham Woods. He accepted a post as lecturer in the Department of Zoology, University of Durham, in 1968.

He remained at Durham for the rest of his career, attaining the rank of Professor in 1987. He also served as Head of the Department of Biological Sciences at Durham from 1990 to 1994. His research work at Durham focused primarily on the ecology and conservation of waders in estuarine waters. He published two books, *Shorebird and Large Waterbird Conservation* (Commission of the European Community, 1983) and *Coastal Waders and Waterfowl in Winter* (Cambridge University Press, 1985). He supervised more than 30 Ph.D. students while at Durham. The BOU awarded him the Godman-Salvin Medal in 1996 for his "key role in this range of ornithological studies and their application and the training of students to go on to further both of these elements."[4] He died in 2001 at the age of 64.

T. ROYAMA (1966)

Tom Royama joined the EGI in October 1962 with funding from the British Council. He had done research on the great tit in Japan, and he joined the team working on the species in Wytham Woods. He used cameras located at the backs of specially designed nest boxes to study the composition of the diet fed to the nestlings. He successfully defended his thesis, *The Breeding Biology of the Great Tit, Parus major, with Reference to Food*, even though, as David Lack wrote, "The thesis of T. Royama on the searching images and insect prey of the Great Tit included facts contradicting three hypotheses earlier put forward by, respectively, his supervisor and each of his examiners (D. Lack, R. A. Hinde, G. C. Varley)."[5] He left the EGI in August 1966 to take a research post at the Forest Research Laboratory of the Canadian Forestry Service in Quebec. He subsequently moved to the Maritimes Forest Research Center in New Brunswick.

Royama continued to work with the Canadian Forestry Service for the remainder of his career, working particularly on the population dynamics of defoliating lepidopteran larvae, including those of the spruce budworm and the white-marked tussock moth. He authored *Analytical Population Dynamics* (Chapman & Hall, 1992), and he received the Gold Medal for outstanding achievement in Canadian entomology from the Entomological Society of Canada in 1994.

URIEL N. SAFRIEL (1967)

Uriel Safriel had never left his native Israel before he accepted a British Council fellowship to pursue a doctorate in England. He had completed his M.Sci. degree in zoology at the Hebrew University of Jerusalem in 1964 and was interested in pursuing a doctorate in animal behavior. After receiving the British Council fellowship, he contacted Niko Tinbergen about the possibility of working with him in Oxford. Tinbergen responded that his group was already full and suggested that he contact David Lack at the EGI. Lack replied to Safriel's inquiry by offering him a post in the EGI and stating that he would be working with birds in the field, and more specifically on the island of Skokholm. Both of these prospects excited Safriel, and despite the fact that he had virtually no background in ecology, he accepted a post as a research student at the EGI, arriving in October 1964.

Safriel worked on the oystercatcher on Skokholm and completed a thesis entitled, *Population and Food Study of the Oystercatcher*. After completion of his doctorate, he accepted a postdoctoral position at the University of Michigan and worked on breeding sandpipers at Point Barrow, Alaska. He then accepted a teaching position at the Hebrew University of Jerusalem in 1969, rising to become a Professor in Ecology, and retired from there in 2009. From 1988 to 1991, he served as Chief Scientist and Director of Israel's Nature Reserves Authority while on loan from the university. From 2002 until 2009, he also served as Professor in Ecology at the Jacob Blaustien Institute for Desert Research.

Safriel's research interests have ranged far from birds. He has worked extensively with intertidal gastropods and has also focused on desert ecology and problems of desertification. In recent years, he has been involved in research on global climate change. He also supervised more than 40 M.Sci. and Ph.D. students at Hebrew University.

R. S. BAILEY (1967)

Roger Bailey arrived at the EGI in October 1961 after completing a B.A. degree. A year later, he was appointed a research student and received a Nuffield Foundation fellowship to serve as marine ornithologist on the *RRS Discovery III*. The newly constructed ship was the principal British vessel participating in an international program of exploration of the Indian Ocean. After spending 18 months aboard *Discovery III*, he returned to the EGI to write his thesis, *The Distribution and Ecology of Seabirds in the Tropical Western Indian Ocean*.

Bailey left the EGI in September 1965 to take a position at the BTO. Two years later, he moved to the Marine Laboratory of the Department of Agriculture and Fisheries for Scotland in Torry, Aberdeen. There he began to work on the assessment of marine fisheries stocks as well as continuing his studies of seabirds. In the early 1990s, he moved to Copenhagen to work for the International Committee for the Exploitation of the Seas. He spent 5 years there advising various ministries on setting catch limits for marine fisheries. He was forced to retire in 1998 because of illness, and he died in 1999.

DEREK A. SCOTT (1970)

Derek Scott joined the EGI as a research student in July 1966, after completing his undergraduate degree in zoology at St. John's College, Oxford, and he immediately left for the island of Skokholm to begin collecting data on breeding storm petrels. He had developed an intense interest in birds while growing up in Yorkshire, and he had hoped to pursue a doctorate under David Lack's supervision since the first time he heard Lack lecture on ornithology during his undergraduate years. He completed his D.Phil. thesis, entitled *The Breeding Biology of the Storm Petrel* Hydrobates pelagicus, in 1970.

After completion of his doctorate, Scott accepted a position as Advisor in Ornithology to the Iran Department of the Environment in Tehran, and he served in that position for 6 years. He then returned to the British Isles and began a consultancy career that has continued into the present. His clients have included the World Bank, the Food and Agriculture Organization and the Development Program of the

United Nations, the World Wide Fund for Nature, the International Waterfowl and Wetlands Research Bureau, the IUCN-World Conservation Union, Wetlands International, and BirdLife International, among many others. His consultancy work has centered on the conservation of wetlands and aquatic birds, focused primarily on the Near East, Africa, South America, and Southeast Asia. He has also served as leader of numerous bird tours to South America, Africa, and Southeast Asia.

M. P. L. FOGDEN (1970)

When Michael Fogden arrived at St. John's College, Oxford, in 1960 to read zoology, he already had an intense interest in birds. He therefore spent some of his time "hanging around" the EGI during his undergraduate years, and after his first year as an undergraduate he was encouraged by Niko Tinbergen to participate in an Oxford-sponsored expedition to the Kiunga Archipelago off the north coast of Kenya. Following completion of the baccalaureate in 1963, he went to Borneo for 3 years to study the seasonality and population dynamics of rainforest birds. In late 1966, he obtained an appointment as a research student in the EGI to write up his research for the D.Phil., entitled *Some Aspects of the Ecology of Bird Communities in Sarawak*. He obtained a Ford Foundation Research Fellowship in 1968 to study the ecology and physiology of resident and migrant warblers in the Queen Elizabeth National Park in Uganda, spending much of the next 4 years there. In 1971, he married Patricia Marshall, who held a Ph.D. in zoology from the University of London.

Michael and Patricia followed in the footsteps of two of Michael's student predecessors at the EGI who had joined with their spouses in becoming a highly productive team: David and Barbara Snow and Philip and Myrtle Ashmole. From 1973 to 1978, the Fogdens worked on range management problems in the Chihuahuan and Sonoran deserts of Mexico for the Centre for Overseas Pest Research of the British Ministry for Overseas Development. They resigned their posts with the Centre in 1978 to become freelance writers and photographers, establishing their base of operations in Monteverde, Costa Rica. They have co-authored several books, including *Animals and Their Colours* (Eurobook, 1974), *The Resplendent Quetzal* (Green Mountain, 1996), and *Hummingbirds of Costa Rica* (Zona Tropical, 2005).

ONE THAT GOT AWAY

John Krebs, the son of Sir Hans Krebs, read zoology at Pembroke College, Oxford, completing his B.Sci. in 1966. His father, a biochemist, was a winner of the Nobel Prize in Physiology or Medicine (1963) who had fled Nazi Germany in 1933 when his university appointment there was terminated. He had come to Oxford in 1954 as Whitley Professor of Biochemistry, and he was also a Fellow of Trinity College. Most of his work was in intermediary metabolism, including the Krebs or citric acid cycle.

John elected to pursue his D.Phil. in zoology at Oxford, and at the Jubilee Conference in Bird Biology in January 1997, he described the events leading to his decision of whom to work with as his supervisor. He first scheduled an appointment with David Lack and spent the morning talking with him about his proposed doctoral research, looking at the potential role of territorial behavior in regulating the population density of great tits in Wytham Woods. After more than an hour discussing his ideas, David and he went for the usual coffee with other members of the EGI. When they walked in, David introduced him by saying, "This is John Krebs. He will be coming to the EGI this fall, but he doesn't know yet what he will be working on." John left the coffee and made an appointment with Niko Tinbergen that afternoon and decided to work with him instead of David. The timing of Krebs' proposed study could hardly have been less propitious. V. C. Wynne-Edwards had recently published his book, *Animal Dispersion in Relation to Social Behaviour*, in which he had identified avian territorial behavior as a prime example of how group selection could operate to maintain populations below levels at which resource depletion and density-dependent mortality would regulate them. Lack was engaged in a spirited, if not heated, debate on the question (see Chapter 9). An alternative explanation for Lack's response, offered by Krebs himself, was that Lack was used to having students who were interested in working on a particular species, with the questions flowing from the biology of the species, not vice-versa.

Krebs' doctoral research did focus on territorial behavior in great tits in Wytham Woods, and it resulted in several significant findings. During the final year of his doctoral program, he joined the EGI as a departmental demonstrator. In 1970, he was appointed Assistant Professor of Ecology at the University of British Columbia, a post he held until 1973, when he returned to Great Britain to become a lecturer in zoology at University College of North Wales, Bangor. In 1975, he returned to Oxford and the EGI as a university demonstrator in zoology. He was appointed Royal Society Professor of Zoology at Oxford in 1988 and held that post until 2005. He also served as chief executive of the National Environment Research Council from 1994 to 1999, and from 2000 to 2005 he was chairman of the U.K. Food Standards Agency. In 2005, he was appointed Principal of Jesus College, Oxford, a post he still held in 2012.

Krebs has received numerous awards for his scientific achievements and public service. He was elected a Fellow of the Royal Society in 1984 and a Foreign Honorary Member of the National Academy of Sciences (USA) in 2004. He was also elected a

Foreign Member of both the American Academy of Arts and Sciences and the American Philosophical Society. He has received numerous awards and medals, including the Croonian Medal of the Royal Society in 2004, and at least 15 honorary doctorates from British and Canadian universities. He was knighted in 1999 and was named Baron Krebs, first Lord of Wythm, in 2007.

ONE WHO FAILED

Like Bill Sladen before him, Bill Bourne had completed a medical education before coming to the EGI. As an undergraduate at Christ's College, Cambridge, he had finished his premedical education in 2 years and read zoology during a third year. He was also active in the Cambridge Bird Club and attended Student Conferences at the EGI. Upon completing his degree at Cambridge in 1951, he moved to St. Bartholomew's Hospital, London, for 4 years of clinical training. In 1955, he was called up by the National Service and sent to Jordan, just in time to be besieged there by three Arab armies during the Suez Canal Crisis of 1956.

Bourne was demobilized from National Service in 1958 and joined the EGI as a field assistant in September of that year. His research was on bird migration in Scotland using radar, but in 1963, his two-volume thesis, *Bird Migration in Scotland Studied by Radar*, was rejected by both Niko Tinbergen and Geoffrey Mathews (the external reader from Cambridge). Bourne had had conflicts with Lack over access to his migration data and its use in publications without Lack's participation, and some attributed his failure to retaliation over these issues.

After this rejection, Bourne returned to the practice of medicine. After 7 years of medical practice in London, he was offered a position as Research Fellow at the University of Aberdeen (Scotland).

These views are speculative, but the restricted ranges of many species cannot easily be attributed to the effects of habitat, food, climate or competitive exclusion by one similar species, and though it is dangerous to argue from ignorance, it is hard to believe that, if many of the existing species were absent, others would not have wider ranges. While this explanation has been introduced here to explain limitations in range, it also explains the limitations in habitat and feeding.

—*Ecological Isolation in Birds*, pp 249–240

11

Ecological Isolation in Birds

IN *ECOLOGICAL ISOLATION in Birds*, David Lack returned to a theme that he first developed in *Darwin's Finches*, the role of competition in structuring ecological communities (see Chapter 3). As in his other synoptic books, much of the data was summarized in a lengthy series of 30 appendices (65 pages), to which he frequently alluded in the text. The underlying premise of his analysis of the global avifauna was the "competitive exclusion principle," which he summarized in the opening sentence of the book: "Two species of animals can coexist in the same area only if they differ in ecology."[1]

The body of the book was devoted to an analysis of the global avifauna to determine how birds belonging to the same genus and living in the same area are ecologically segregated. (The choice of the genus as the unit of comparison was based on the fact that members of the same genus are typically very similar, both morphologically and ecologically.) Therefore the question becomes, "How are members of the same genus living in the same general area separated ecologically?" The three means Lack recognized as important in segregating species ecologically were (1) differences in geographical range, (2) differences in habitat for species that share the same range, and (3) differences in feeding for species that occupy the same habitat. For the last category, he used size differences as a surrogate if data were lacking on their actual foods:

Finally, where two species living in the same habitat differ markedly in size, and particularly in size of beak, they have been assumed to differ in their feeding, as this has been established in many instances, and there are no known exceptions.[2]

Being initially skeptical about Lack's claim that such differences in feeding had been adequately established for many species, especially when the size differences were small, I counted the number of times he had used this assumption in the book as evidence of ecological segregation (113 times).[3]

Lack began his analysis close to home, looking at the ecological relationships of the tits (Paridae) living in Europe, and then proceeded to examine this group throughout the world. Chapters followed on the nuthatches, two groups living in human-altered environments (swallows and finches), and migratory species living in two geographical locations. He then considered the remaining avifaunas of several large regions, with chapters devoted to Europe, North America, Africa, other tropical areas, archipelagoes, and remote islands. Although he identified numerous examples of segregation by both geographical range and feeding, habitat differences tended to prevail in most groups: "A difference in habitat is much the commonest means of ecological isolation between congeneric species in continental passerine birds and is widespread in most other groups."[4]

Reviews of *Ecological Isolation in Birds* were mixed. For instance, Robert Ricklefs, writing in *Science*, praised the book for pulling together data from widely scattered sources but decried the "all too rare touches of synthesis."[5] A similar tone was struck by another reviewer:

> The book is difficult because of the beguiling simplicity of this doctrine [ecological isolation]. The reader continually wonders *why* a particular pattern of ecological isolation is observed in a certain genus or location, and *how* it came to evolve instead of some alternative pattern. These considerations are scarcely mentioned.[6]

The remainder of this chapter deals with Lack's relationships with mentors and colleagues. An analog of the competitive exclusion principle may apply to the nature of some of these relationships.

SIR JULIAN HUXLEY

Although David's life and work were influenced by a number of eminent contemporaries—most notably Ernst Mayr, Niko Tinbergen, and Reg Moreau—Julian Huxley, serving as his unofficial mentor, played a central role at many of the most pivotal junctures in David's professional life. Huxley was not only a pioneering ethologist and committed evolutionary biologist but also a philosopher-statesman. Although he never served in any formal sense as Lack's teacher, it is hard to overestimate his influence on Lack's development as a scientist.

FIGURE 17: Julian Huxley talking with David and Elizabeth Lack at the 1950 International Ornithological Congress in Uppsala. (Photography courtesy of Lack family.)

Julian Sorrell Huxley, born in 1887, had an unusual pedigree in both science and literature. His grandfather was "Darwin's bulldog," Thomas Henry Huxley, and his mother was the niece of the poet of Oxford's "dreaming spires," Matthew Arnold. His younger brother, Aldous, became a celebrated novelist, perhaps best known for *Brave New World* (Chatto & Windus, 1932), and a younger half-brother, Andrew, shared the Nobel Prize in Physiology or Medicine with Alan Hodgkin in 1963. Julian attended Eton and then read zoology at Balliol College, Oxford, but he also won the Newdigate Prize for Poetry for *Holyrood* in 1908 (his great-uncle had won the prize in 1843).

After graduation, Julian remained at Balliol as a lecturer in zoology (1910–1912) and then left to become Professor of Biology at the Rice Institute in Houston, Texas (1913–1916). He returned to England during World War I to serve in the Army Intelligence Corps. After the war, he returned to Oxford as Senior Demonstrator of

Zoology and Fellow of New College, Oxford (1919–1925), before ending as Professor of Zoology at King's College, London (1925–1927). He then left formal academic life to pursue research and writing full-time.

It is difficult to depict in a few words the impact of Huxley's scientific contributions. At the time of his death in 1974, Peter Medawar described him as "one of the foremost biologists of the 20th century."[7] His scientific contributions ranged from physiology and experimental embryology to evolution and animal behavior. He was a co-founder of both the Society for Experimental Biology (1923) and the Association for the Study of Animal Behaviour (1936). He was a major contributor to the neo-Darwinian revolution, particularly with two books, *The New Systematics* (Clarendon Press, 1940) and *Evolution: The Modern Synthesis* (George Allen & Unwin, 1942).

After World War II, Huxley was deeply involved in the establishment of the United Nations Education, Scientific and Cultural Organization (UNESCO), with many crediting the addition of the "S" in its charter to his persuasive influence. He served as UNESCO's first Director-General from 1946 to 1948.

Huxley was elected a Fellow of the Royal Society in 1938, and he won its prestigious Darwin Medal in 1956. He was knighted in 1958.

David's first introduction to Huxley was by reading his studies of avian courtship behavior, which were lent to him during his last term at Gresham's School, particularly Huxley's classic *The Courtship Habits of the Great Crested Grebe* (Cambridge University Press, 1913) and a similar study of the red-throated diver (loon). Their first recorded personal encounter was in May 1933, when David invited Huxley to speak to the Cambridge Bird Club. It was at that time that Huxley began to act as David's unofficial supervisor, a role he played for a number of promising young scientists.

Huxley's first action on David's behalf was in obtaining the teaching position at Dartington Hall School. As described in Chapter 2, Huxley spoke at Dartington Hall, and it was presumably at that time that he learned that the school had an opening for a science mentor. He recommended Lack and also suggested to David that he apply for the position, which he did. David taught at Dartington Hall from 1933 to 1940.

The second instance of Huxley's influence was his suggestion that David become acquainted with tropical environments, and also that he contact Reg Moreau as a potential guide. David took this suggestion to heart and as a result took the trip to Tanganyika during his first summer holiday from his teaching post at Dartington Hall (see Chapter 2).

Huxley was secretary of the Zoological Society of London in 1937, when David was seeking funding to support an expedition to the Galapagos Islands during a sabbatical from Dartington Hall. Huxley was influential in securing grants from both the Royal Society and the Zoological Society of London, thereby providing most of the funding for the expedition. The ultimate product of the expedition, *Darwin's Finches*, proved to be the fulcrum on which the future of Lack's professional life turned.

Huxley apparently supported Lack for the directorship of the Edward Grey Institute of Field Ornithology (EGI) in 1945 despite the fact that the other candidate, James Fisher, had listed Huxley as one of his two referees (see Chapter 5). When David was elected a Fellow of the Royal Society in 1951, primarily on the basis of his work on the Galapagos finches, Huxley was again a major supporter of his candidacy.

ERNST MAYR

Ernst Mayr, one of the leading ornithologists and evolutionary biologists of the 20th century, was born in southern Bavaria in July 1904. He began to work for Erwin Stressemann at the Natural History Museum in Berlin during holidays from studying medicine at the University of Greifwald. Stressemann convinced Mayr to study for a Ph.D. in zoology and ornithology after completion of his medical education in 1925. Mayr earned the doctorate in 1926 and thereafter began working full-time at the Museum in Berlin.

In 1929, Mayr joined an expedition to New Guinea jointly sponsored by the Berlin Museum, Lord Walter Rothschild's Natural History Museum at Tring, and the American Museum of Natural History (AMNH) of New York. Afterward, the AMNH persuaded him to join their Whitney South Sea expedition to the Solomon Islands and then invited him to come to New York on a 1-year appointment to work with the expedition's collections. Mayr arrived in New York in January 1931, and he never returned to the Berlin Museum. He continued to work at the AMNH until 1953, when he was appointed Alexander Agassiz Professor of Zoology at Harvard University's Museum of Comparative Zoology, a post he held until his retirement in 1975. He remained active as Emeritus Professor of Zoology until shortly before his death in 2005 at the age of 100.

Mayr's most salient scientific contributions were in the field of evolutionary biology. His development of the biological species concept and the theory of allopatric speciation, both in the early 1940s, constituted a major contribution to the emerging neo-Darwinian Synthesis. In many respects, his *Systematics and the Origin of Species from the Viewpoint of a Zoologist* (Columbia University Press) and Huxley's *Evolution: The Modern Synthesis*, both published in 1942, represented the culmination of a process that began with R. A. Fisher's *The Genetical Theory of Natural Selection* (Clarendon Press) in 1930 and included significant contributions from J. B. S. Haldane, Theodosius Dobzhansky, G. Ledyard Stebbins, Sewell Wright, and others.

After retirement, Mayr's interests turned increasingly to the history and philosophy of biology. *The Growth of Biological Thought* (Belknap Press, 1982) and *Toward a New Philosophy of Biology* (Harvard University Press, 1988) are representative of these expanded interests.

Mayr received numerous honors for his many achievements, including election to membership in the National Academy of Sciences (USA) in 1954 and as a Fellow in both the American Academy of Arts and Sciences (1954) and the American Association for the Advancement of Science (1958). He received the International Balzan Prize in 1983 and the Darwin Medal from the Royal Society of London in 1984. Recognition of his work in the history and philosophy of biology came in the form of the Salton Medal (History of Science) in 1986 and the Benjamin Franklin Medal (American Philosophical Society) in 1996. He received the President's National Medal of Science in 1969. He was also awarded an honorary D.Sci. degree from Oxford University during the International Ornithological Congress held in Oxford in 1966, no doubt with the strong support of David Lack, who was president of that Congress (see Chapter 8).

David met Mayr while visiting the AMNH on the last day of his trip to the United States in 1935. David's journal account of the meeting was brief: "Then called on Dr. Mayr, who was the perfect host, showed me the magnificently housed Tring collection [Lord Rothschild's collection of birds, which had recently been acquired by the AMNH], and had a long discussion on New Guinea birds and altitudinal distribution."[8] Mayr later recalled their first meeting: "We liked each other at once because our interests and our ways of thinking were so similar."[9]

The next meeting between the two occurred in 1939, when David was en route back to England following his Galapagos expedition and his sojourn at the California Academy of Sciences (see Chapter 3). He lodged with the Mayrs in their New Jersey home for the duration of his stay in New York. As noted in Chapter 3, the two were together when they heard the news that their respective nations were at war.

Mayr was a member of the planning committee for the Princeton Conference on Genetics, Paleontology, and Evolution held in January 1947 (see Chapter 5), and he was no doubt involved in identifying David as a potential speaker at the conference. In a March 1946 letter to David, Mayr wrote,

> If it should work out I hope that you can discuss the ecological aspects of speciation. There is no other field about which there is more confusion and

FIGURE 18: Ernst Mayr and David Lack at D.Sci. ceremony for Mayr in Oxford during the 1966 International Ornithological Congress. (Photograph courtesy of Lack family.)

muddled thinking. The fault is, of course, partly with the geneticists who have never clearly defined the genetic conditions for the establishment of reproductive isolation and ecological divergence.[10]

Lack did participate in the conference, and he subsequently visited several major American universities, at most of which he gave seminars (see Chapter 5).

In the years that followed, the two friends met frequently at international meetings and corresponded regularly. David often asked Mayr for comments on manuscripts on which he was working, including all or part of *Darwin's Finches, The Natural Regulation of Animal Numbers*, and *Island Biology Illustrated by the Land Birds of Jamaica*—the last even though his principal thesis challenged some of Mayr's own ideas. David did consistently support Mayr's theory of speciation, asserting that geographical isolation is necessary for species formation in birds.

NIKO TINBERGEN

Nikolas Tinbergen was born at the Hague, the Netherlands, on April 15, 1907, and grew up there in a family that included three brothers and a sister. He attended Leiden University, completing undergraduate studies in biology and a Ph.D. in biology by 1932. He married Elisabeth (Lies) Rutten that same year, and the couple left shortly thereafter for a 15-month meteorological expedition to eastern Greenland, where they lived with an Inuit tribe. After returning from Greenland, Niko worked as a laboratory instructor in biology at Leiden University, where he designed laboratory experiments with animals that could be readily maintained in captivity, including the three-spined stickleback. The stickleback later became a major research subject in the nascent field of ethology. More importantly, Tinbergen was simultaneously working in gull colonies near the Hague. In 1940, he was appointed to a lectureship in behavior at Leiden.

World War II interrupted Niko's academic career. He was imprisoned for 2 years in a "hostage camp" by the Dutch SS for failing to support the "Nazification" of Leiden University, which included the removal of Jewish professors. After being released from the camp, he returned to his position at the university, and in 1947 he was appointed Professor of Animal Behavior.

Shortly after the end of the war, Niko organized a symposium on "factors determining distribution in animals" at Leiden and invited several prominent scientists, including the newly appointed director of the EGI, David Lack, to participate (see Chapter 5). After the February 1946 symposium, Niko traveled to Oxford with Lack, and he also visited William Thorpe at Cambridge. He had decided that he wanted to study and teach ethology in an English-speaking country, and both Lack and Thorpe were hoping that he would choose their institution. Niko also visited Ernst Mayr in New York during a 3-month visit to Canada and the United States that began later that year. Mayr was interested in luring Tinbergen to the United States. In the end, with further encouragement from David Lack, Niko returned to Oxford in February 1949 to negotiate with the head of the Zoology Department,

Alister Hardy. He accepted a position as demonstrator in the department and moved with his family to Oxford in September 1949.

Upon his arrival in Oxford, Tinbergen established the Animal Behaviour Research Group (ABRG) in the Zoology Department. The unit attracted numerous outstanding D.Phil. students including, during Niko's early years in Oxford, Aubrey Manning, Desmond Morris (author of *The Naked Ape*), and Mike Cullen, and during later years, John Krebs, Richard Dawkins, and Hans Kruuk (Niko's biographer). Niko preferred being in the field to being in the university, and much of his time and that of his students was spent working in gull colonies at Ravenglass or Walney on the Irish Sea. Although he had received an appointment at Merton College on his arrival in Oxford, he took little interest in college life. He was appointed Reader in Zoology in 1960 and Professor of Zoology in 1966, holding the latter post until his retirement in 1974. He was also appointed Fellow of Wolfson College, Oxford, in 1966.

Tinbergen wrote several books in addition to his many scientific papers. Among the more important were *The Study of Instinct* (Clarendon, 1951), *Social Behaviour in Animals: With Special Reference to Vertebrates* (Methuen, 1953), and *The Herring Gull's World: A Study in the Social Behaviour of Birds* (Collins, 1953). After retirement from the university, he and Lies turned their attention to behavioral studies of childhood autism, and in 1983 they co-authored the book, *"Autistic" Children: New Hope for a Cure* (George Allen & Unwin).

In 1973, Tinbergen shared the Nobel Prize in Physiology or Medicine with Konrad Lorenz and Karl von Frisch. The three, the only field-oriented biologists ever to win a Nobel Prize, were recognized for their major contributions to the establishment of the field of ethology, based primarily on work done during the 1930s. Niko's older brother, Jan, had shared the first Nobel Prize in Economics with the Norwegian Regnar Frisch in 1969. Niko's 1973 prize made them the first brothers to become Nobel laureates.

Niko received numerous other awards for his accomplishments. He was elected a Fellow of the Royal Society in 1962. He was awarded the Godman-Salvin Medal by the British Ornithologists' Union (BOU) in 1969, and both the Jan Swammerdam Medal (Dutch Academy of Science and Arts) and the Wilhelm Bölsche-Medal in 1973.

David apparently initiated the relationship with Niko in early 1940 by sending him reprints of some of his papers. Niko responded with a three-and-a-half page letter, in the first paragraph of which he said, "I studied your paper of the Robin with great interest and have discovered, that you are not only an ecologist but also a first class behaviour student."[11] World War II interrupted their correspondence, but shortly after the war ended they were again in contact. As noted earlier, Tinbergen organized a conference at Leiden University which Lack attended by invitation (also see Chapter 5). Niko entrusted him with funds that were deposited in a British bank, and David used the money to purchase and send items that the Tinbergens were unable to obtain in post-war Holland. One of the more crucial purchases was a bicycle to replace the one that had been "requisitioned" by fleeing German soldiers.

The Tinbergens and the Lacks formed a close and enduring personal bond during the Tinbergens' first few months in Oxford. David and Elizabeth had been married only 2 months when the Tinbergens arrived in late September 1949. For 3 months the Tinbergens lived in a converted pub, the Worcester Arms, in Islip south of Oxford, but in December they moved to a house on the Banbury Road in North Oxford, close to the Lacks' flat in Park Town (see Chapter 6). The two couples were together often during the next 2 years until the Lacks moved to Boars Hill.

In many respects, Niko and David were unlikely friends. Niko was very outgoing and fun-loving, a true "people person." David, on the other hand, was quite shy and often socially awkward and usually seemed to be anything but fun-loving. Another difference involved the relationship between work and home. Niko integrated the two, hosting Friday night seminars for members of the ABRG at his home until the group became too large. David kept his home and workplace quite segregated, particularly after the move to Boars Hill. The two men did share a number of important characteristics, however. Both loved the out-of-doors and had a particular passion for birds. They also had a strong commitment in their respective disciplines to evolutionary explanation based on natural selection operating at the level of the individual. Both were workaholics. So, despite the differences, they remained close friends until David's death.

One might well ask why, with this mutual interest in field work on birds and their close friendship, the EGI and the ABRG did not collaborate in their research efforts. There is no definitive answer to that question, but the physical separation of the two units may have played a role. The ABRG was located in the Zoology Building next door to the University Museum, whereas the EGI was located at the Botanic Garden. Doctoral students in the two units often were in contact with each other, but there was no shared research program. At least one EGI student, Bryan Nelson, left the EGI after a year to work with Mike Cullen in the ABRG, primarily because he decided to work on the gannet rather than the European blackbird. Niko often attended the annual Student Conferences in Bird Biology held by the EGI, giving plenary talks at some of them. He no doubt used the conferences, as Lack did, for "talent scouting" for promising doctoral students.

Mayr and Tinbergen shared one major bit of influence in David's career. Their strong research programs in avian evolutionary biology and avian behavior, respectively, encouraged David to choose avian ecology as his major focus of research when he became Director of the EGI in 1945 (see Chapter 5).

REG MOREAU

Reginald Ernest Moreau, whom Lack said "might well be reckoned the first African ecologist,"[12] was one of the last persons who might have been identified as a candidate for such acclaim. Born in 1897 near London, where his father was a member of the London Stock Exchange, his prospects diminished dramatically when his father was seriously injured in a freak accident when Reg was 10 years old. He was struck by the open door of a passing express train while standing on the platform, and a subsequent nervous breakdown left him permanently incapacitated. After completing grammar school, Reg sat for the Executive Class Examination of the Home Civil Service, and at the age of 17 he became an auditor for the War Office. In 1920, he applied for and received a transfer to the Cairo station of the Army Audit Department. He recalled, "I was just 23, about as ignorant and uninformed as I had been half-a-dozen years before."[13] It was in Cairo that his passion for birds developed— and, quite by accident, he met an English visitor interested in birds who soon became Winnie Moreau. Reg's interest in birds led him to become self-educated in ornithology and related subjects, and he published several papers on Egyptian birds and contributed two chapters to Richard Meinertzhagen's *Nicoll's Birds of Egypt* (Hugh Rees, 1930). Just as he was completing his fourth tour of duty in Egypt and, as he described it, "the War Office was all set to pull me back into the chummy embrace of home service,"[14] serendipity provided him the opportunity to take a temporary position with the Colonial Service at the new Agricultural Research Center in Amani, Tanganyika. The Moreaus arrived at Amani in March 1928, and by 1934 Moreau had made significant contributions to African ornithology as well as tropical ecology.

David met Moreau when the latter hosted him for a month in August-September 1934 at the Agricultural Research Center at Amani (see Chapter 2). Moreau was forced to retire from his position at Amani in 1947 due to an eye infection that threatened his sight, and David offered him a half-time position as research officer in the EGI. He had also been appointed editor of *Ibis* in 1947, a position he held for 13 years, and he proved to be a consummate editor. He retired from the EGI post in August 1964 but continued to come into the Institute regularly until he left Oxford for Herefordshire in 1966. He died in May 1970.

Most of Moreau's research during his time at the EGI was devoted to the birds of Africa, including the species that wintered there. His major work on the latter subject, *Palearctic African Bird Migration Systems* (Academic Press, 1972), was published posthumously. He wrote *The Bird Faunas of Africa and its Islands* (Academic Press, 1966) and co-authored, with B. Patricia Hall, *An Atlas of Speciation in African Passerine Birds* (British Museum 1970). He also wrote *The Departed Village: Berrick Salome at the Turn of the Century* (Oxford University Press, 1968).

Moreau received numerous awards in recognition of his contributions to African ornithology. He was elected an Honorary Fellow of the American Ornithologists' Union in 1949, and in 1962 he received the second Stamford Raffles Award "for contributions to ornithology." The University of Oxford awarded him an honorary Master of Arts degree in December 1951. He served as president of the BOU from 1960 to 1965 and received the society's most prestigious award, the Godman-Salvin Medal, in 1966.

The friendship that David forged with Moreau when he visited Tanganyika in 1934 evolved into a close and productive working relationship after Reg began working at the EGI in 1947. In his capacity as the editor of *Ibis*, Moreau, along with David, contributed substantially to a transformation of scientific ornithology in Great Britain during the middle of the 20th century.[15] David made extensive use of Moreau's editorial skills, which no doubt improved the clarity of his writing. Indeed, in a footnote at the end of the Acknowledgements in *Ecological Isolation in Birds*, David wrote:

> When this book was in press, R. E. Moreau died. I owe him an immense debt, because he read in manuscript every book and paper that I wrote over a period of more than twenty years, and he was unexcelled as a critic.[16]

Perhaps as importantly, Reg helped shape the culture of the EGI by serving as a foil for its Director. He was the only member of staff with sufficient seniority and experience to challenge or chide David in the tea-time discussions, especially when David made some imperious pronouncement. He was also able to interject some humor. One D.Phil. student concluded that the EGI would have been "pretty tedious without Reg."[17]

GEORGE COPLEY VARLEY

David Lack and George Varley were born in the same year, and both arrived in Cambridge in 1929, Varley to read natural sciences in Sidney Sussex College. He graduated in 1933, winning the Frank Smart Prize in Zoology, and remained at Cambridge to pursue his Ph.D. He earned his doctorate in 1935 with a study of the knapweed gallfly. During his graduate studies, he became acquainted with the work of the pioneering Australian entomologist, A. J. Nicholson.

Following the completion of his doctorate, Varley was appointed Research Fellow at Sidney Sussex College, a post he held for 3 years before spending a year as a professor at the University of California. In 1939, he returned to Cambridge as a University Demonstrator and Curator of Insects. He remained in that post until 1945, although during most of World War II he worked in the Army Operational Research Group (AORG). It was there that he reconnected with David Lack, whom he had known during his undergraduate years at Cambridge.

After the war, Varley accepted a position as a reader in entomology at King's College, Newcastle, which he held for 3 years. In 1948 he was appointed to the Hope Professorship in Entomology of the University of Oxford (a post that was much coveted by the great Oxford ecological geneticist and entomologist, E. B. Ford, but denied him because of his widely known proclivity for male undergraduates). Varley held the post until his retirement in 1978. In Oxford he once again renewed his relationship with David Lack, and his principal research program, on the winter moth in Wytham Woods, dove-tailed nicely with the tit population studies in Wytham conducted by the EGI (see Chapter 5). Winter moth caterpillars were a major food source for the tits. Varley co-authored (with G. R. Gradwell and M. P. Hassell) the important work, *Insect Population Ecology: An Analytical Approach* (Blackwell, 1973).

Varley married Margaret "Peggy" Brown in 1955. She also had a Ph.D. from Cambridge and was an internationally known ichthyologist. They lived in North Oxford until George died in 1983. Peggy died in 2009.

George was also an avid bird watcher, and it was probably at the Cambridge Bird Club that he and David became acquainted during their undergraduate years. Their second contact was in the AORG during the war, where both worked in radar. It was actually Varley who first observed that the "angels" appearing on the radar screens were actually flying birds (see Chapter 2). They co-authored a paper in *Nature* describing their wartime discovery of radar detection of flying birds. These observations led ultimately to radar studies of migration and other movements by birds conducted by the EGI in the early 1960s (see Chapter 8).

Varley made a second significant contribution to Lack's subsequent career during the war. He introduced David to A. J. Nicholson's concept of density-dependent population regulation, an idea that figured prominently in David's book, *The Natural Regulation of Animal Numbers* (see Chapter 5). In addition, David credited wartime discussions with Varley for helping to sharpen his grasp of competitive exclusion, which reshaped his understanding of his findings on the Galapagos finches (see Chapter 3). As noted in Chapter 6, George served as David's best man when he and Elizabeth were married in July 1949, and the two men remained friends until David's death.

W. H. THORPE

William Homan Thorpe was born April 1, 1902, in Hastings, East Sussex. As a boy, he developed strong interests in natural history and music, and after learning that one could earn a living as an agricultural entomologist, he went up to Jesus College, Cambridge, in 1921 to read agriculture. During his undergraduate years, Thorpe and his good friend, Edward Armstrong, were among the founding members of the Cambridge Ornithological Club (later renamed the Cambridge Bird Club during David Lack's tenure as president). After completing his undergraduate degree in Agricultural Science, he received a Rockefeller Foundation fellowship to study biological pest control at the University of California. He returned to Cambridge in 1929 and completed his Ph.D. that same year. From 1929 to 1932, he worked as an entomologist at the Imperial Bureau of Entomology at Farnham Royal before returning to Cambridge as a lecturer in entomology and Fellow of Jesus College. In 1939, he and his wife, Mary, visited the Agricultural Research Center at Amani, Tanganyika, where they were hosted by Reg and Winnie Moreau. Thorpe was amazed at the Moreaus' knowledge of tropical birds.

Thorpe had had a long interest in birds, but it was the reading of several papers by Konrad Lorenz that converted him from an insect physiologist to an ornithologist and ethologist. He decided to study the development of song learning in birds and pioneered the use of tape recording and sound spectography in the analysis of bird song. He was instrumental in convincing the Zoology Department at Cambridge to establish the Ornithological Field Station in the village of Madingley, 6 miles from Cambridge, in 1950, and he was appointed as its director. Robert Hinde, who was completing his D.Phil. under David Lack in the EGI, became curator of the fledgling institution. The two of them, along with their students, quickly expanded its spectrum of research to include mammals, and today it is the world-class research unit known as the Sub-Department of Animal Behaviour, with research programs that include neuroscience and human cognition.

Thorpe was elected the first president of the Association for the Study of Animal Behaviour in 1948. He also served as president of the BOU from 1955 to 1960, a period that included the centennial celebration of the organization. He received the prestigious Godman-Salvin Medal from the BOU in 1968. In 1951, he was elected a Fellow of the Royal Society "for his researches on insect physiology and animal behaviour."[18]

David met Thorpe during his final undergraduate year at Cambridge when the latter returned as reader in entomology. Thorpe recalled later that Lack "was not fired by the zoology teaching in Cambridge. . . . However the David Lack of the [Cambridge] Bird Club was an entirely different creature, absolutely dedicated, bubbling over with enthusiasm and outstandingly learned and experienced for his age."[19] Thorpe apparently suggested that Lack read J. B. S. Haldane's *The Causes of Evolution* (Longmans, Green, 1932), which, along with R. A. Fisher's "The Evolution of Dominance" in *Biological Reviews* (1931), Lack described as "the first non-ornithological works to excite my zoological imagination."[20]

By the time Lack assumed the directorship of the EGI after the war, Thorpe was in the process of changing his research focus from insect physiology to the study of bird song. Thorpe was one of the two references Lack provided to the Oxford Committee for Ornithology when he applied for the directorship. The two men interacted frequently during the years that followed, particularly on the plans for an ornithological field station at Madingley. As noted earlier, Robert Hinde, Lack's first D.Phil. student, became the curator of the field station when it was established in 1950. Both Lack and Thorpe were elected Fellows of the Royal Society in 1951.

After David's conversion to Christianity in 1948 (see Chapter 7), he and Thorpe had something else in common. Thorpe belonged to the Society of Friends. It was Thorpe who recommended David to John Wren-Lewis as a worthy substitute when he declined Wren-Lewis's invitation to participate in the lecture series, "Modern Cosmology and Christian Thought," held at St. Anne's House, London, in the fall of 1953 (see Chapter 7).

BERNARD W. TUCKER

"With the death of Bernard William Tucker on 19th December 1950 after a long illness, British ornithology loses its central figure."[21] So wrote David Lack to express his admiration for a man who had been a close colleague and friend since Lack's arrival in Oxford in October 1945. Born in Hertfordshire in the first month of the new century, Tucker had first come to Oxford in 1919 as an undergraduate at Magdalen College to read zoology, receiving First Class honors in 1923. He then received a scholarship to the Stazione Zoologica in Naples, France, where he spent 2 years before being appointed a Demonstrator in the Zoological Laboratory at Cambridge. In 1926, he returned to Oxford as University Demonstrator in Zoology and Comparative Anatomy, remaining at Oxford until his death. At the time of his death, he was a reader in ornithology.

As an undergraduate at Oxford, Tucker had helped to found the Oxford Ornithological Society, the first such society at a British university. Upon returning to Oxford in 1926, he resumed his affiliation with the society and, along with Max Nicholson, played a major role in creating the Oxford Bird Census, the entity that eventually led to the formation of the British Trust for Ornithology (BTO) and the

EGI (see Chapter 5). He continued to play such a key part in the development of these two entities that Lack asserted at the time of Tucker's death:

> [H]e more than anyone else, was responsible for gaining University support for the Institute [EGI]. It may be recalled that, through over-emphasis by Professor [Alfred] Newton, and because of their relatively uniform morphology, birds had been effectively banished from the Animal Kingdom by orthodox zoologists for some thirty years. It was a bold step for a British university to recognize field ornithology, which zoologists tended to regard as the domain of the dilettante amateur. Indeed, that this recognition no longer seems absurd is largely due to the Oxford example, and hence to Tucker.[22]

This was a fitting tribute for an esteemed colleague and close friend.

CHARLES ELTON

One of the enduring mysteries of the history of 20th century ecology is that two of its pioneering practitioners worked literally side by side as directors of the Bureau of Animal Population (BAP) and the EGI, yet had virtually no productive interaction. In fact, famously, the door separating the BAP and the EGI in their Botanic Garden quarters was kept locked.

Charles Sutherland Elton, widely recognized as the father of animal ecology, was born in Withington, Manchester, in 1900. His grandfather, the Rev. Charles Elton (1820–1887), had at one time been the Headmaster of Gresham's School, Holt, where David Lack would later study. Elton, however, attended a day school at Liverpool College before going up to New College, Oxford, in the fall of 1919 to read zoology. His tutor was Julian Huxley, who managed to guide the young Elton through the very uninteresting (to Charles) comparative anatomy and descriptive embryology that comprised the bulk of the zoology curriculum of the day. When he graduated in 1922, the warden of New College, William Spooner (who famously lent his name to malapropisms such as his announcement of the next hymn as *Kinquering Congs their Titles Take*), wrote Charles such a confusing letter that he had to bicycle to North Oxford to consult with the sub-warden to ascertain that New College was offering him a 2-year stipend. He was to spend his entire professional life at Oxford.

During his undergraduate days, Charles participated in the first Oxford expedition, in 1921, to the Arctic island of Spitsbergen. The expedition was organized by

Oxford biologists and consisted of 18 men from Oxford, Cambridge, and several other institutions. Charles served as Huxley's field assistant, and his observations of the Arctic communities to which he was exposed had a profound influence on his pioneering ideas in community ecology. During the next 10 years, he participated in two other expeditions to Spitsbergen and one to Norwegian Lapland.

In 1926, Huxley asked Charles to write a book on animal ecology for a series that he was editing, "Textbooks of Animal Ecology." The resultant work, entitled *Animal Ecology* (Macmillan, 1927), was written by Elton in only 85 days. It laid the framework not only for animal ecology but for community ecology as well. In the book, Elton introduced such novel concepts as the pyramid of numbers, the functional niche, characteristic species, food chains, species richness, and population cycles. The book established a new discipline, animal ecology, and provided the major rationale for considering Elton the father of animal ecology.

Elton was appointed a university demonstrator in 1929, and in 1932 he established the BAP as a research institute devoted to the study of animal populations. He served as Director of the BAP from 1932 until his retirement in 1967. The primary focus of research at the BAP was the regulation of mammal populations, particularly cyclic mammals such as lemmings and voles, and their predators, whose populations were also usually cyclic. During World War II, the institute devoted most of its effort to the control of rodent pests threatening the food supply.

After the war, Charles instituted an ecological survey of the habitats and organisms of Wytham Woods near Oxford. The Wytham Biological Survey generated an enormous amount of data, all of which were encoded on punch cards in the early days of the computer age. In addition to *Animal Ecology*, Elton authored several other books, one of which, *The Ecology of Invasions by Animals and Plants* (Chapman and Hall, 1958), proved to be another pioneering work. It is now by far his most frequently cited work, a status it reached only after his death in 1991. He retired from the BAP in 1967 but remained active until shortly before his death.

David initiated a correspondence with Elton while he was still a student at Cambridge, after one of his own Arctic expeditions. Both men participated in the symposium on the implications of Gause's principle of competitive exclusion at the annual meeting of the British Ecological Society in March 1944. But it was not until Lack became Director of the EGI in 1945 that the two were in frequent contact. Not only were their respective institutes literally next-door neighbors in the St. Hugh's garden and at the Oxford Botanic Garden, but for most of the first 3 years of their marriage, David and Elizabeth lived not 100 yards from Charles and his wife, the poet Joy Scovell, in Park Town. Nevertheless, from the time that the BAP and the EGI moved from the garden at St. Hugh's to the Oxford Botanic Garden until Elton retired in 1967, the door between the two institutes was locked. The BAP was phased out of existence upon Elton's retirement, being incorporated into the Animal Ecology Research Group (AERG) in the Department of Zoology. One of the first acts taken by John Phillipson when he assumed leadership of the AERG in 1967 was to unlock the door at the Botanic Garden.

Why did Elton and Lack fail to develop a fruitful partnership? Speculation continues to this day about the answer to that question. Students at their respective institutes interacted freely and productively, but their directors seemed always to maintain a cordial aloofness with each other.

Students and staff of both the BAP and the EGI referred to their respective directors as "the Boss," but only among themselves and not to their directors' faces. Although the students and staff of the institutes generally worked harmoniously, the coolness between Lack and Elton affected their attitudes toward the leader of the other organization. Scuttlebutt around the BAP in the mid-1950s included the belief that Lack had gotten a Third in zoology at Cambridge, and that was why he had found it necessary to go into teaching in a public school. There was also a rumor among the BAP personnel that much of *The Life of the Robin* had been pirated from another source (presumably Birchett, who had done similar studies on the robin in Ireland) (see Chapter 2).

There are few clues as to the feelings of the two protagonists in the conflict. Elton was by all descriptions a courtly but rather shy individual who rarely expressed strong opinions about others. One hint about his attitude toward Lack comes from an incident reported by Dennis Chitty in his autobiographical *Do Lemmings Commit Suicide? Beautiful Hypotheses and Ugly Facts* (Oxford University Press, 1996). Chitty arrived at the BAP in 1935, completed a D.Phil. under Elton's supervision, and remained at the Bureau until 1961, when he accepted an appointment at the University of British Columbia. In June 1956, Elton wrote a strongly-worded, critical letter to Dennis, probably related to the latter's statement made at a conference in Munich earlier that spring that continent-wide population cycles of a species were not synchronous. In the letter, Elton drew negative comparisons with two other scientists (whose names were left blank in *Do Lemmings Commit Suicide?*). One of the references was to Lack: "In the last few years you have shown a strong tendency to do what [David Lack] does, select the data that fit and under-rate those that don't."[23] This statement makes it clear that Elton was critical of Lack's perceived selective use of evidence to support his own hypotheses.

In his turn, Lack was also critical of some of Elton's science. He expressed considerable skepticism about the amount of time devoted to the Wytham Biological Survey, believing that the potential utility of the resultant mass of data in answering meaningful ecological questions was very limited and could not possibly justify the immense expenditure of time and energy. This opinion reflected a fundamental difference in the scientific approaches of the two men. Lack was essentially a population ecologist, and he used a hypothetico-deductive approach to address specific questions based on evolutionary principles. Elton was first and foremost a systems ecologist and believed that meaningful ecological answers could not be achieved without an understanding of the ecosystem within which the relevant organisms were embedded. Robert MacArthur spent a postdoctoral year in the EGI in 1957–1958, and he is reported to have expressed his opinion of the two men in the following couplet: "David Lack has perception. Charles Elton has perspective."

The two directors also differed markedly in personality. Elton was reserved and very shy, whereas Lack, although fundamentally a shy person, was more aggressive, sometimes even brash—characteristics that were frequently interpreted as abrasive. Joy Scovell was also extremely shy and reserved, which meant that the Eltons and Lacks did not interact socially.

Some specific incidents may also have contributed to the coolness between the two men. Lack was named a Fellow of the Royal Society in 1951; Elton was also nominated for the honor, but he did not receive it until 2 years later. Some have suggested that Elton considered the selection of the much younger Lack something of an injustice and an affront.

One of the few instances in which Elton was observed to express anger publicly occurred 12 years later. J. W. S. Pringle replaced Alister Hardy as Head of the Zoology Department in 1961, 2 years before the latter's mandatory retirement age, so that Pringle could oversee the development of the new Zoology Building (see Chapter 12). Hardy continued on for 2 years as Head of the Department of Zoological Field Studies, which included the BAP and the EGI. When Hardy retired in 1963, Pringle announced at a departmental meeting that he was appointing David to fill the post of Head of Zoological Field Studies. Elton, who was at the meeting, objected strenuously and somewhat intemperately to the appointment of the younger and less senior Lack to the post. David was not at the meeting, and it is not known how he reacted to the now twice-spurned Elton's objections. The position had little more than a titular function. Some redress of the perceived injustices came in 1970 when Elton received the Darwin Medal from the Royal Society, 2 years before Lack's selection.

A professional principle of competitive exclusion may actually have been at work in the lack of cooperation between Lack and Elton. The primary interest of both men was in identifying factors that regulate animal populations. And, as in the case of many Oxford dons, one's closest intellectual colleague is often one's greatest rival.

MAX NICHOLSON

Another person with whom David frequently failed to "see eye-to-eye" was Max Nicholson. Edward Max Nicholson was born in Ireland but moved to England while still young. He attended Sedbergh School, Cambria, but was unable to complete his studies due to lack of family financial support. He developed a passion for studying birds early in life, and when he went to work in London writing guide books he was encouraged to apply for a scholarship to Oxford by the editor of *The Times*. His application was successful, and he went up to Hertford College, Oxford, to read history at the age of 22; shortly afterward, his book, *How Birds Live: A Brief Account of Bird Life in the Light of Modern Observation* (Williams and Norgate, 1927), was published. He demonstrated his organizational abilities early by convincing the university to establish the Oxford University Exploration Club, the first such club at a British university. He was also active in the Oxford Ornithological Society and, along with Bernard Tucker, helped to establish the Oxford Bird Census, which eventually developed into the EGI and the BTO (see Chapter 5).

After graduating from Oxford, Nicholson became assistant editor of the *Weekend Review*, the first of numerous jobs in several fields, including government and conservation. He served in the Ministry of War Transport during World War II, and after serving as head of the Office of the Lord President of the Council from 1945 to 1952, he became Director-General of the Nature Conservancy, a post he held for 14 years. He was one of 16 signatories (along with Julian Huxley and Peter Scott) of the Morges Manifesto, which led to the founding of the World Wildlife Fund in 1961.

Despite his myriad other responsibilities, Nicholson continued to be deeply committed to and involved in the BTO, including serving as its chairman from 1947 to 1949. He also remained committed to his original vision of the relationship of the BTO to the EGI (see Chapter 5). This vision included the idea that one of the principal responsibilities of the EGI was to provide direction and training for large numbers of amateurs collecting data on birds. Soon after he became Director of the EGI in 1945, David Lack began to express suspicions about the quality of data generated by amateurs, and he was disinclined to make a large commitment to their training. Instead, he saw his role as primarily the development of a strong field-based research program within the EGI itself. It was at this juncture that he and Nicholson parted ways. Nicholson had supported James Fisher's candidacy for the directorship over Lack's, perhaps because he perceived that David would move the institute in the direction he did. Relations between the two men became very strained, and the close connection between the EGI and the BTO gradually dissolved (see Chapter 5). Although they maintained a superficial cordiality in the years that followed, they were never close.

Max Nicholson was a man with both remarkable organizational ability and creative vision. For his many contributions to the field of ornithology as an amateur, the BOU awarded him its highest honor, the Godman-Salvin Medal, in 1962. He died in 2003 at the age of 98.

G. EVELYN HUTCHINSON

Ironically, another pioneering ecologist of the 20th century, G. Evelyn Hutchinson, completed his public school education at Gresham's School just 3 years before Lack's arrival. Hutchinson, whose father was an outstanding mineralogist at Cambridge, returned to Cambridge to enroll at Emmanuel College for undergraduate and doctoral training and then taught at Yale University for most of his illustrious academic career. He is widely recognized as the father of modern limnology and was also a major contributor to the fledgling field of ecology. He was elected a member of the National Academy of Sciences (USA) and a Fellow of both the American Association for the Advancement of Science and the American Academy of Arts and Sciences. He received the Presidential National Medal of Science in 1991 (the year of his death in London). The latter award, perhaps the most prestigious science recognition in the United States, has been given to 441 scientists in all fields since its creation in 1959.

Hutchinson's connections with David Lack extended well beyond their common background at Gresham's. It is unclear when they first met. Hutchinson had completed his studies at Cambridge before Lack arrived as an undergraduate in 1929. Hutchinson did invite Lack to Yale following the Princeton conference in January 1947 (see Chapter 5). Another prominent ecologist, Robert H. MacArthur, spent a postdoctoral year at the EGI, immediately after completing his Ph.D. under Hutchinson at Yale in 1957 (see Chapter 9). And Prof. Chris Perrins, a doctoral student of Lack's and his successor as Director of the EGI, vividly remembered giving Hutchinson a ride on the back of his motorcycle to view the swifts in the tower at the University Museum during one of the latter's visits to Oxford.

RICHARD MEINERTZHAGEN

Although the name of an acclaimed American ornithologist best known for his studies of West Indian birds provided Ian Fleming with the inspiration for the name of his fictional British spy 007, an ornithologist closer to home actually prefigured some of James Bond's exploits. Richard Meinertzhagen—soldier, spy, explorer, and ornithologist—was one of the most enigmatic characters in 20th century ornithology.

He was born March 3, 1878, into a wealthy and well-connected family. Educated at Harrow and the University of Gottingen, he joined the army rather than take a position in the family's banking business. He served primarily in Africa and Asia during his military and espionage careers, rising to the rank of Colonel, and chronicled his exploits in several memoirs.

As an adolescent Meinertzhagen had developed a strong interest in natural history, particularly in birds, and in all his travels he collected birds, eventually amassing a large collection of specimens. He wrote *Nicoll's Birds of Egypt* (Hugh Rees, 1930), to which Reg Moreau contributed (as described earlier). He also published numerous papers in ornithological journals. In most respects, Meinertzhagen seemed to embody the classic explorer-ornithologist who contributed to the museum-based, geographical ornithology described by Kristin Johnson.[24] He died in 1967.

Two decades after his death, Meinertzhagen's story began to unravel. Careful detective work by Alan Knox, Patricia Rasmussen, and Robert Prys-Jones revealed that Meinertzhagen had stolen numerous bird specimens from major museums and had fabricated data on the labels of many of his own specimens. These revelations resulted in a general discrediting of his scientific contributions. But what of his military and spying exploits? A recent book by Brian Garfield, *The Mystery of Meinertzhagen: The Life and Legend of a Colossal Fraud* (Potomac Books, 2007), suggested that his other supposed accomplishments were as suspect as his ornithological contributions.

Even more ominously, suspicions began to emerge about the "accidental" death of his second wife, the ornithologist Anne Constance Meinertzhagen, in July 1928. H. F. Witherby described the accident in Anne's obituary in *British Birds*:

Col. And Mrs. Meinertzhagen had been engaged in revolver target-practice and while returning to the house Mrs. Meinertzhagen was examining her revolver in the belief that it was empty; this, however, was not so and her husband, who was walking ahead, was startled by an explosion and turning round saw his wife fall; and she was found to be dead.[25]

There was no inquest at the time, and Anne's death was long considered to be a tragic accident.

Members of the BOU apparently harbored no such suspicions about Meinertzhagen. The BOU awarded him its highest honor, the Godman-Salvin Medal, in 1951.

David's first contact with Meinertzhagen came in the form of a congratulatory letter, following publication of *The Life of the Robin*, in which Meinertzhagen asked several questions about the robin, including whether David knew where robins roosted. Correspondence between the two continued sporadically thereafter, with Meinertzhagen expressing in one letter his apparent dislike of the robin:

> You have no doubt remarked on the huge size of the Robin's eye, almost twice the size of that of the Dunnock. As you know he is crepuscular in habit and I have seen him pounce from a stance in moonlight. He is certainly the last to bed and earliest up which fits in with his greed and aggression. He has all the characters of Mussolini and I know of no bird less suited to association with the birth of the Messiah.[26]

Clearly, Meinertzhagen's and Lack's views on the robin were quite divergent.

J. B. S. HALDANE

"He has an inordinate fondness for beetles." Thus J. B. S. Haldane is reputed to have responded to a query posed by the Bishop of Oxford at a college dinner: "So what, Prof. Haldane, have your researches taught us recently about the nature of God?" One can only speculate about the weighty tenor with which the question was asked, but the answer reveals much more about the brilliant but iconoclastic and acerbic mathematician-biologist who contributed significantly to the formulation of the Modern Synthesis than it does about the nature of God. The story may be apocryphal, but it illustrates the conclusion of his colleague C. D. Darlington in his review of Ronald Clark's Haldane biography, *The Life of J. B. S. Haldane* (Hodder and Stoughton, 1968):

> [H]e craved for applause. And he was driven by his unforeseen limitations to seek it more and more in the public rather than the scientific world. In this way, he was given a myth with which his style and figure well accorded. And quite early this myth took control of his life.[27]

John Burdon Sanderson Haldane, son of the well-known Oxford physiologist, John Scott Haldane, was born November 5, 1892, in Oxford. Educated at the Dragon School (only a stone's throw from the Haldane Estate in North Oxford that is now the location of Wolfson College) and Eton College, he went up to New College, Oxford, to read mathematics, graduating with First Class honors in 1914. After serving in World War I, he returned to Oxford to pursue graduate studies in biochemistry, and in 1922 he assumed the post of reader in biochemistry at Trinity College, Cambridge, which he held until 1933. He then moved to University College, London, as Professor of Genetics from 1933 until 1937, when he was appointed Professor of Biometry. He resigned this position 20 years later, in protest against the British and French response to the Suez Canal Crisis, and emigrated to India, where he took a post as research professor at the Indian Statistical Institute in Calcutta, from which he retired in 1961. He died in India in 1964.

Haldane's scientific contributions were wide-ranging, including major contributions in biochemistry, population genetics, physiology, and the philosophy of science in addition to his role in articulating the modern evolutionary synthesis already mentioned. He is widely credited with coining the word "clone" in a 1963 lecture, deriving it from the Greek word meaning twig. He received a number of scientific awards. He was elected a Fellow of the Royal Society in 1932 and received both the Royal Society's Darwin Medal (1952) and the Linnean Society's Darwin-Wallace Medal (1958). He also received the Kimber Genetics Award from the National Academy of Sciences (USA) in 1961.

As indicated by the excerpt from C. D. Darlington, however, Haldane's interests exceeded the bounds of science. While he was at Cambridge, Trinity College was the epicenter of Communist activism in the university (see Chapter 1). Haldane visited the Soviet Union in 1931, along with Julian Huxley, J. D. Bernal, John Cockcroft, and other scientists, and he gradually embraced Communism during the years that followed, joining the Communist Party in 1937. He actively supported the Republican cause during the Spanish Civil War, traveling to Spain to aid in developing effective air-raid shelters. He resigned from the Party in 1950, in part to protest the rise of Lysenkoism in Soviet genetics. During his time in the Communist Party, he contributed regularly to the *Daily Worker*, including serving on its editorial board from 1940 to 1950.

As noted earlier, David, at William Thorpe's suggestion, read Haldane's *The Causes of Evolution* during his final year at Cambridge. It is not known whether David met Haldane before leaving Cambridge, but it is unlikely given that the latter was a reader in biochemistry, a field far from David's interests.

They did begin a correspondence later, particularly after David became Director of the EGI. In December 1946, they sailed aboard the same ship en route to the Princeton conference (see Chapter 5). Haldane made the arrangements for their passage. Their last correspondence came just a year before Haldane's death. David was in the process of writing *Population Studies of Birds*, and he asked Haldane for permission to make reference to his "Pangloss's theorem," a request with which Haldane complied in a letter dated April 25, 1963. Haldane also identified four other versions of the law in that letter, including

The Hermit's (of Prague) theorem, "That which is, is." For example "A Pi meson is in an excited state, therefore a Pi meson is." In theology this is called the ontological argument. This is almost a corollary of the Bellman's theorem. If I make enough statements about the Devil, the Categorical Imperative, Innate Releasing Mechanisms, or what you will, these gradually acquire existence.[28]

David did make reference to Pangloss's theorem in his critique of Wynne-Edwards' group selection theory in *Population Studies of Birds*,[29] and the theorem later received a much wider audience in Stephen Gould and Richard Lewontin's critique of the contemporary adaptationist program, which they referred to as the Panglossian paradigm.[30]

MICK SOUTHERN

Henry Neville Southern was born in 1908 and while still in school developed a strong interest in natural history, particularly birds. In 1927, however, he went up to Queen's College, Oxford, to read classics, graduating in 1931. He remained interested in natural history, and while an undergraduate he spent a great deal of time photographing birds. He also wrote *Close-ups of Birds* (Hutchinson, 1932), which was published the year after his graduation. Little wonder, therefore, that he returned to Queen's in 1935 to read zoology. He graduated with First Class honors and began working in Charles Elton's BAP in 1939. During World War II, he worked on the BAP's study of mammalian pests of agriculture and food stocks, but after the war he began his parallel, long-term studies of two rodents, the wood mouse and the bank vole, and their main predator, the tawny owl, in Wytham Woods.

Southern made numerous contributions to several scientific societies. He served as the first editor of the BTO's journal, *Bird Study*, from 1954 to 1960, and also edited the *Journal of Animal Ecology* from 1968 to 1975. He served as president of the British Ecological Society from 1968 to 1970, and of the Mammal Society from 1974 to 1980. The BOU awarded him its Union Medal in 1971, and the Mammal Society presented him its Silver Medal in 1974. He received a D.Sci. degree from Oxford in 1972. Mick died in 1986.

Given Mick's interest in birds and his cheerful disposition and obvious enthusiasm for his work, it is not surprising that he was David Lack's most consistent contact in the BAP. Mick was invited by the Oxford Committee for Ornithology to apply for the directorship of the EGI in 1945 but declined with the caveat, "I only wish to apply for this post in the event of Mr. David Lack's candidature lapsing" (see Chapter 5).

After David was appointed Director of the EGI, he and Mick spent 3 weeks on the island of Tenerife in the Canary Islands, studying its avifauna (see Chapter 5). Mick also took the black-and-white photographs of the swifts breeding in the Museum Tower that appeared in *Swifts in a Tower* (see Chapter 6). David invited Mick to be one of the senior speakers at the 1954 Student Conference, when the topic was The Population Problem.

V. C. WYNNE-EDWARDS

Vero Copner Wynne-Edwards was born In Leeds on July 4, 1906. After attending Rugby, he went up to New College, Oxford, in 1924 to read zoology. His first tutor at New College, Julian Huxley, left for King's College, London, after a year, and Charles Elton succeeded him as Wynne-Edwards' tutor. During his undergraduate years, Wynne (as he was called by friends) was active in the Oxford Ornithological Society and participated in the establishment of the Oxford Bird Census, the precursor of the BTO and the EGI. He graduated with First Class honors in 1927 and became a senior scholar at New College from 1927 to 1929. He then spent a year as lecturer in zoology at Bristol University before accepting a position as an assistant professor at McGill University in Montreal.

During the 15 years that he spent at McGill, Wynne worked in a variety of areas. He studied seabirds in the North Atlantic, making important observations on both their distribution and their movements. He also worked on freshwater fishes, particularly those of the St. Lawrence River and its tributaries, and he developed an interest in montane plants. He won two Walker Prizes (Boston Society of Natural History) and was elected a Fellow of the Royal Society of Canada.

Wynne-Edwards was named Regnus Professor in Zoology at the University of Aberdeen in 1946, after Alister Hardy had vacated the professorship to become Head of the Zoology Department at Oxford (see Chapter 5). Wynne continued in that post until his retirement in 1974. He published his *magnum opus, Animal Dispersion in Relation to Social Behaviour*, in 1962, and the ensuing debate between Lack and Wynne-Edwards about the latter's proposal regarding the role of group selection in the regulation of animal populations is described in Chapter 9.

Animal Dispersion in Relation to Social Behaviour, although controversial, stimulated much discussion and new research. Wynne-Edwards received numerous honors and awards for his many contributions. These included election as a Fellow of the Royal Society of Edinburgh (1950) and the Royal Society of London (FRS, 1970). He was awarded a D.Sci. degree from Oxford, and he received the Godman-Salvin Medal from the BOU in 1977. He died in 1997 at the age of 90. At the end of an obituary, two of his closest colleagues wrote, "[H]is portrait must hang in the gallery of the "greats", beside those of Lack, Elton, Huxley and Hardy, and within view of Darwin's."[31]

ROBERT H. MACARTHUR

Robert Helmer MacArthur was born in Toronto, Canada, in 1930 and later moved with his family to Marlboro, Vermont, where his father was a genetics professor at Marlboro College. After completing his undergraduate degree at Marlboro College, Robert earned a master's degree in mathematics from Brown University. He then enrolled in a Ph.D. program at Yale University to study ecology under G. Evelyn Hutchinson. His doctoral research on niche-partitioning by several warbler species in the genus *Dendroica* breeding in spruce forests in Maine quickly became a classic when it was published in *Ecology* in 1958.[32] The study was clearly inspired in part by Lack's work on the Galapagos finches (see Chapter 3).

Robert MacArthur arrived at the EGI in October 1967 for a 1-year postdoctoral fellowship in the EGI with David Lack. He, his wife Betsy, and their 18-month-old son took up residency on Headington Hill. During their year in Oxford, their second child was born in the home, with a doctor, a midwife, and Robert in attendance. Betsy recalled that Elizabeth Lack would invite the children and her to the Lack home on Boars Hill for tea, a welcome interlude from the isolation she experienced during Robert's days at the EGI.

Robert returned to the United States in 1958 to become Assistant Professor of Biology at the University of Pennsylvania. He moved to Princeton University in 1965 as Professor of Biology, a post he held until his death in 1972. He co-authored (with E. O. Wilson) the extremely influential book, *The Theory of Island Biogeography* (Princeton University Press, 1967), and he also wrote another highly acclaimed book, *Geographical Ecology: Patterns in the Distribution of Species* (Harper & Row, 1972).

He was elected to the National Academy of Sciences (USA) in 1969 and was awarded the Eminent Ecologist Award by the Ecological Society of America (ESA) in 1973. In 1983, the ESA created the Robert H. MacArthur Award, which is given biannually to an established ecologist in mid-career for meritorious contributions to ecology.

Robert and David remained close friends after Robert's year in Oxford, and they corresponded with each other often. Robert and Betsy also stayed with the Lacks when they visited Oxford in 1969. In his final book, *Island Biology*, David developed his own ideas about the numbers of bird species on oceanic islands, challenging some of the assumptions and conclusions in MacArthur and Wilson's *Theory of Island Biogeography* (see Chapter 13).

Such instances have given rise to fruitless arguments as to the correct name to use. But as has often happened in science something that was first regarded merely as a nuisance has become the basis of a new theory. For many years, there was great dispute as to how new species might originate, but enough is now known to say that, at least for birds, the only populations with hereditary differences less than those which separate different species are geographical races or subspecies of the same species, and they show every gradation from forms just separable from each other by average measurements to forms as different as full species. The conclusion is clear that, at least in birds, species arise from geographically isolated populations.

—*Evolution Illustrated by Waterfowl*, pp 39–40

12

Evolution Illustrated by Waterfowl

TWO OF DAVID Lack's books were published posthumously. The first, a slender introduction to evolution aimed at a general audience, was entitled *Evolution Illustrated by Wildfowl* and was published by Blackwell Scientific Publications in conjunction with The Wildfowl Trust of Slimbridge, Gloucestershire (now The Wildfowl and Wetlands Trust). The Wildfowl Trust was founded in 1946 by Peter Scott, son of the famed Antarctic explorer, Robert Falcon Scott. Peter was a contemporary of Lack's during their undergraduate years at Cambridge, and they remained friends in the years that followed. Lack frequently included a field trip to Slimbridge for students participating in the annual Student Conferences in Bird Biology (see Chapter 5), and Scott usually made a presentation. Scott, who was knighted in 1973, was still director of The Wildfowl Trust at the time of the publication of *Evolution Illustrated by Wildfowl*.

Evolution Illustrated by Waterfowl was aimed at A-level or undergraduate students and informed amateur naturalists. The text was organized into 23 chapters, which were completed in only 83 richly illustrated pages. Robert Gillmor again did the excellent pen-and-ink drawings used in the text. The language was free of technical jargon, and when technical terms were required for clarity, they were usually defined in the text as well as in a glossary at the end of the book (Appendix 2). Appendix 1 was a list of the 147 waterfowl species of the world organized by tribe and including scientific and common names for each species.[1]

In the first chapter, entitled "The Problem," Lack wrote:

> How did this diversity of forms come into existence, and why do they differ
> from each other? This problem is, in miniature, that of all animal life, and the
> waterfowl have been used in this booklet to show the main principles of animal
> classification and evolution.[2]

The next few chapters dealt with methods of classification, describing problems
that develop in identifying species, genera, and higher-order relationships among the
swans, geese, and ducks of the world. Four chapters dealt with subspecies, the final
one describing the transition from subspecies to species, in which Lack concluded
that, "at least in birds, new species arise from geographically isolated populations."[3]

Subsequent chapters dealt with a wide range of subjects including hybridization,
sexual selection, interspecific competition, adaptive radiation, adaptations for
breeding, and molt migration ("a specialty of waterfowl"[4]). In the final chapter, after
mentioning the importance of the Galapagos finches in Darwin's thinking, Lack
summarized:

> The wildfowl likewise provide a model of evolution, but in their case it is
> necessary to choose examples from many different regions. The chief aim of
> this book has been to demonstrate the successive stages from simple mutations
> to an adaptive radiation.[5]

Evolution Illustrated by Waterfowl remains a very readable introduction to evolution by means of natural selection.

THE EDWARD GREY INSTITUTE, 1967–1973

Following the hosting of the International Ornithological Congress in July 1966 and
the departure of Reg Moreau to Hertfordshire a month later, the Edward Grey Institute of Field Ornithology (EGI) entered into a period of transitions that culminated
tragically with the death of its Director in March 1973. These transitions included
Moreau's death in May 1970, David Lack's 10-month sabbatical to the West Indies in
1970–1971, and another move, this time from its picturesque and intimate quarters
in the Oxford Botanic Garden to a massive new concrete edifice on the corner of
South Parks and St. Cross Roads, the new Zoology Building.

Departures of D.Phil. students from the early 1960s opened opportunities for a
new crop of students. The two students who had been sponsored by the British

Council both left the year they completed their degrees, Tom Royama in 1966 to take a post with the Forest Research Laboratory in Quebec, Canada, and Uriel Safriel in 1967 for a 1-year postdoctoral fellowship at the University of Michigan before returning to Israel. Ian Newton, who had remained in the EGI as a research officer after completing his D.Phil. in 1964, left in October 1967 to take a position with the Nature Conservancy in Scotland. Peter Evans, who had remained in the EGI as a departmental demonstrator for 3 years after completing his D.Phil., left in October 1968 to take a position as lecturer in the Department of Zoology at Durham University. Only Chris Perrins remained in the EGI, as a senior research officer.

Several new D.Phil. students arrived in the EGI, and in the "annual report" for the 2 years that included his 10-month sabbatical in the West Indies, Lack related that the focus of research in the EGI had shifted to two principal emphases, physiological ecology of birds (under Chris Perrins' leadership) and the ecology of tropical birds. The concentration on tropical birds included Lack's own interest in the birds of the West Indies but also the interests of two senior visitors in the EGI, Peter Ward and Michael Fogden. The focus of the physiological ecology research was on the energetics of egg production and chick development, with David Dawson (from New Zealand) and Luc Schifferli (from Switzerland) studying clutch size and the energetics of egg production in the house sparrow and Raymond O'Connor working on chick development patterns in the great tit, common swift, and house sparrow.

In January 1970, David attended the annual meeting of the Serengeti Research Institute (SRI) in Seronera, Tanzania, where he had been asked to serve on the scientific advisory committee. He and his oldest son, Peter, flew to Nairobi on an overnight flight from London, covering in less than 12 hours essentially the same route that had required 7 days for David to traverse in 1934 (see Chapter 2). After the meeting, Hugh Lamprey, Director of the SRI, flew David and Peter to the Nuffield Unit of Tropical Animal Ecology (now the Uganda Institute of Ecology) at Lake Katwe, Uganda, near Queen Elizabeth National Park. They spent a week with Michael and Patricia Fogden, who were working there. One of the highlights of their stay was discovering the largest barn swallow roost any of them had ever seen, while searching a swamp for chimpanzees, which they did not see. The roost numbered at least in seven figures. Michael then drove them about 600 miles to Lake Nakuru, where they spent 2 days before going on to Lake Naivasha and then returning to Nairobi. From there, David flew back to England, and Peter returned to the SRI to begin a 7-month position as field assistant. Peter had completed his studies at Bryanston, and David had arranged for the position in a move reminiscent of his own father's placement of him at the Senckenberg Museum in Frankfurt am Main in 1929 (see Chapter 1).

David, who at the time also served on the research committees of the British Trust for Ornithology, the Royal Society for the Protection of Birds, and The Wildfowl Trust, continued to serve on the science advisory committee of the SRI and returned for another meeting in January 1972. On that trip, he also visited the Tsavo National Park in Kenya.

STUDENT CONFERENCES

Student Conferences in Bird Biology were convened each January in St. Hugh's College, with the 27th annual conference being held just 2 months before Lack's death. Conference themes over the years included Evolution Below the Level of the Species, Annual Cycle, Ecological Isolation, Principles of Conservation, Social Behaviour and Ecology, Tropical Ecology, and Birds and Insects. Plenary speakers continued to maintain the high standard established at the beginning of the Student Conferences. Michael Fogden presented three plenary talks, and Peter Evans, Ian Newton, Peter Ward, and Michael Harris each presented two. Professors P. M. Sheppard, J. R. Baker, and Richard Southwood (who was to replace John Pringle as Head of the Zoology Department at Oxford in 1979) also served as plenary speakers.

Student attendance had grown to about 60 annually, and many of the participants read research papers or made poster presentations. David noted that the 1967 conference, "Evolution Below the Level of the Species," was somewhat flat but that the next conference, "Annual Cycle," generated considerably more interest.

A SABBATICAL IN JAMAICA

In June 1969, David was invited to a symposium sponsored by the Association for Tropical Biology on Adaptive Aspects of Insular Evolution. The symposium was held in Mayaguez, Puerto Rico, and Lack presented a paper outlining some of his own ideas on insular avifaunas.[6] He also used the trip to make preparations for his anticipated sabbatical the following year. Before flying to Puerto Rico, he arranged to spend time with a pair of young ornithologists, Cameron and Angela Kay Kepler, who were studying the parrots and todies of the archipelago. He stayed with them for 2 weeks at their home in the moist tropical forests of Sierra de Luquillo and also accompanied them to visit the lowland dry forests of Guanica. There his lifelong interest in nightjars, dating back to his schooldays at Gresham's (see Chapter 1), manifested itself in his particular interest in seeing the Puerto Rican whip-poor-will.

David also visited Jamaica on this trip. In Kingston, Jamaica, he met with Prof. Ivan Goodbody, Professor of Zoology at the University of the West Indies, to discuss arrangements for the sabbatical. Goodbody had briefly been a student at the EGI in 1949.

David was appointed Royal Society Visiting Professor at the University of the West Indies for the 1970–1971 academic year, and the family, minus Andrew, who was doing his A-levels at Bryanston, sailed from Southampton on September 30, 1970, aboard the *TSS Camito*, a banana boat bound for Kingston, Jamaica. Some of the excitement and anticipation that the family felt about this 10-month sojourn on an exotic tropical island is reflected in David's description of their arrival in Jamaica:

> Early on the morning of October 12, 1970, a fortnight late owing to a British dock strike, our ship moved slowly into Kingston, which has one of the finest harbours in the world, and as a light mist dispersed, the whole line of the Blue Mountains came into view, forming a magnificent backdrop to the north.[7]

Also accompanying the Lacks were David's principal field assistant, A. W. (Tony) Diamond, and Tony's bride, Liz. Diamond had recently completed his doctorate at the University of Aberdeen and was selected as field assistant after Lindon Cornwallis, one of Lack's D.Phil. students, decided not to take the position. Accompanying the party were two vehicles, the Lacks' personal car and a Land Rover, and about 100 other passengers.

Upon arrival in Jamaica, the Lacks took up residence in a large bungalow in a northern suburb of Kingston about 2 miles from the university. Because the Diamonds did not yet have housing of their own, the newlyweds were obliged to stay with the Lacks for the first few weeks on the island. Tony later recalled[8] that to enter the bathroom, they had to pass through the younger children's bedroom— a not entirely enchanting honeymoon. Eventually Tony and Liz secured their own housing in Irish Town, about halfway up the Blue Mountains northeast of Kingston.

Tony's primary responsibility was to mist-net and band (ring) birds on the island, focusing on endemic Jamaican species but also banding seasonal migrants from North America if they were captured. He suffered from ulcerative colitis and had a serious flare-up of the illness while in Jamaica, which prevented him from being as productive as he had expected to be, and afterward he believed that David was disappointed with his efforts. David expressed no hint of such a disappointment, however. Tony also took dozens of photographs of the birds he captured, both endemics and migrants, which later proved very useful.

Peter also served as a field assistant to his father, and they worked together censusing Jamaican birds and documenting their foraging habits. David invited Kay Kepler, who was still working in Puerto Rico and wished to make observations on the Jamaican

FIGURE 19: David and Elizabeth in Jamaica, 1971. (Photograph courtesy of Lack family.)

tody, to come to Jamaica to accompany them on their censuses. She came and stayed with the Lacks for 2 or 3 weeks in April and May 1971. She described the typical day in the field[9] as beginning with a wake-up call at about 3 a.m. because she, Peter, and David often had to drive some distance to the morning's census site. Peter, on the other hand, recalled the typical morning beginning at about 4 a.m. The long drive was required because of David's unwillingness to camp overnight near the site, a reluctance that may have been based on reservations about camping that he expressed in a letter to Tony Diamond before the trip:

> He [an unnamed Oxford zoologist who had been in Jamaica 2 years earlier], like everyone except in part the one [unidentified] American, says that you should not camp, and further that it is unnecessary to camp. The natives are widely scattered through the countryside and anywhere that you camped you would collect a crowd.[10]

Another reason for not camping was that David did not wish to leave Elizabeth and the two younger children alone at night any more than necessary due to concerns about the high crime rate in Kingston.

No matter the time of their departure for the field, Elizabeth dutifully rose to prepare a lunch for the crew. Kay recalled that on the first such morning, Elizabeth asked if she wanted cheese, lettuce, or tomato on her sandwich, and that when she replied that she wanted all three, Elizabeth made it clear that the "or" meant just that. Kay also recalled that although Elizabeth shopped for fresh bread at the market each day, she didn't use the freshest bread until the last of the previous day's bread was gone. Anyone who has attempted to raise a large family on the rather modest wages of an academic will appreciate the origin of Elizabeth's frugality. Peter described the censusing days as beginning with a 1- to 3-hour drive to the day's site, followed by 3 to 4 hours of censusing and the drive home, sometimes for a late lunch. Lunch was often followed by a short siesta before writing up the day's notes and then an hour's swim in the Olympic-size pool at the University, usually with Elizabeth and the other children (his father rarely took time for the swim). At 6 p.m., Elizabeth served a tasty home-cooked meal, a somewhat challenging task because the oven in the bungalow was not working. David would often read for an hour and then retire promptly at 9 p.m. Kay recalled that David enjoyed using the local names for the Jamaican birds, and the name for the Jamaican tody, Robin Redbreast, must have been a particular delight to him (see Chapter 4). Indeed, in *Island Biology* he referred to it as a "gem."

David visited several other islands in the West Indies, including Barbados (February 1–3), Trinidad (February 3–12 and July 23–27), St. Lucia (March 9–13), St. Vincent

(March 13–17 and July 15–23), Grenada (March 17–21 and August 1–9), Grand Cayman (May 17–20), Puerto Rico (May 29–31), Dominica (May 31–June 4), and Tobago (July 27–August 1). During the February visits to Barbados and Trinidad, David lectured at the local campuses of the University of the West Indies as part of his responsibilities as a visiting professor. Either Peter or Tony accompanied him on most of the visits to the other islands, and the entire family traveled together on the July trips to St. Vincent and Trinidad. From Trinidad, Elizabeth, Peter, Paul, and Catherine flew back to Kingston to catch the boat to England, while Andrew, who had now competed his final year at Bryanston, traveled on to Tobago and Grenada with his father before the two of them flew back to England. Andrew recalled:

> The most vivid time for me was that two weeks at the end, first in Tobago and then in Grenada, my little 'lollipop' for not being there all the rest of the time. It was simply idyllic for me, at least in part because I had my father to myself.... Those two weeks were particularly precious in retrospect because it was the last time I really talked in any depth with him, perhaps the first time in that sort of depth too. It meant, and still means, a huge amount to me, and I find writing this brings much of the emotion back—41 years later.[11]

One rather frightening event that occurred while the two were hiking on Grenada was that David slipped and fell down a precipitous slope beside the ridge path they were following, disappearing completely from view. Andrew helped direct his hand to a small tree near where he landed about 25 feet below the path, and David managed to pull himself up the slope with no serious injury save the loss of his glasses (actually his spectacles rather than his binoculars, which he referred to as his "glasses").

David and Elizabeth returned to the West Indies in June 1972 for some follow-up observations, spending 2 weeks on Dominica doing censuses similar to those done the previous year on Jamaica. It was their first field trip alone together since their second visit to the Pyrenees to study migration in September 1950. David's description of Dominica in his 1972 Christmas letter was ebullient: "A superb island, three-quarters still forest, the last of the West Indies still to be seen as Columbus saw them."[12] His enthusiasm for the island extended to its people, whom he described as "6 feet tall and boisterous, speaking English and a French patois, ... kindliness itself." Just before their trip, David had had X-rays taken as part of an annual examination required by the University, but he did not get the results until his return from the Caribbean. Those results were to change everything.

The data collected on the sabbatical, as well as that from the follow-up visit, served as the basis for his final book, *Island Biology as Illustrated by the Land Birds of*

Jamaica (see Chapter 13). Elizabeth and Peter played major roles after David's death in seeing that the book was published, and Kay Kepler did the line drawings of the birds of Jamaica.

THE MOVE TO "HMS PRINGLE"

In 1961, Sir Alister Hardy retired early as Head of the Department of Zoology to enable his successor, John Pringle, to begin planning for a new Zoology Building and overseeing its construction. As mentioned in Chapter 8, Hardy retained his position as Head of the Department of Zoological Field Studies for 2 more years, until his mandatory retirement at age 67.

The site for the new building was the corner of South Parks Road and St. Cross Road. Leslie Martin, a well-known English architect, was chosen to design the structure. He was a strong proponent of the International Style, one of the early schools of modern architecture, and is perhaps best known for his design of Royal Festival Hall in London. The massive concrete monstrosity that he conceived for the Zoology Building could scarcely be more out of place anywhere than in Oxford, with its dreaming Medieval spires. The building was derisively referred to as "Pringle's Folly," "Pringle's Palace," "Pringle's Cave," or "HMS Pringle" by many members of the department. Its gleaming white concrete exterior aged poorly, fading to a dingy gray. The building, which has housed both the Zoology Department and the Experimental Psychology Department since it was completed, has now been named the Tinbergen Building, after Niko Tinbergen, a member of the Zoology Department from 1949 until 1974 (see Chapter 11). Probably the best thing that can be said about the Tinbergen Building is that the windows on the south side of the building offer a truly magnificent view of Oxford's "dreaming spires." And this view does not include the incongruous "HMS Pringle."

Construction delays resulted in the building's not being ready for occupation until early 1971, while David was on sabbatical in Jamaica. Therefore, Chris Perrins, who was serving as acting director of the EGI during David's absence, was forced to oversee the move from the institute's home in the Botanic Garden to the new quarters. This included moving the contents of David's office to his new office in the southeast corner of Floor D, two stories above the ground-level floor. The Alexander Library of Ornithology was moved to a more expansive space on Floor C, which David remarked would be adequate for 25 years of growth. In fact, it was 40 years before another move took the Alexander Library to a yet bigger space on Floor B. The current space has no outside window, and consequently no spectacular view of old Oxford, but what one does get there is a panoramic view, perhaps unparalleled anywhere, of the literature on the world's birds, an ever-growing legacy of the EGI's four directors, W. B. Alexander, David Lack, Chris Perrins, and Ben Sheldon.

ILLNESS AND DEATH

When David and Elizabeth returned from Dominica in June 1972, his physician called to set up an appointment to discuss a troubling shadow on his lung that had shown up on the X-rays taken during his pre-trip examination. Surgery was performed at Churchill Hospital in Headington to remove what doctors presumed was a tumor in the lung. What they discovered instead was an enlarged lymph node, and further tests revealed that David was suffering from non-Hodgkin's lymphoma, a cancer caused by the uncontrolled proliferation of lymphocytes, a type of white blood cell. His consulting physician recommended radiation therapy, to be followed by chemotherapy. Several forms of non-Hodgkin's lymphoma occur, differing in aggressiveness (slow- or fast-growing) and in the type of lymphocyte involved. Unfortunately, it quickly became apparent that David had a fast-growing form of the disease.

Radiation therapy and chemotherapy in the early 1970s were not as sophisticated in type and targeting as they are today. The treatments often resembled swinging at a fly with a sledge hammer. Consequently, the treatments themselves could have significant negative side-effects on the patient's health. Compounding this problem for David was the fact that the technician administering the radiation treatment irradiated the wrong side of the body on his first attempt, thereby necessitating a second debilitating radiation treatment. The radiation therapy was followed by a course of chemotherapy, again a debilitating experience for the patient. David did not return to the EGI after the chemotherapy treatment began because of fear of infection due to the extremely low white blood cell counts that followed. In a Christmas letter sent in December 1972, he described his illness and said that after the surgery he had suffered from "'operational shock' whatever that is"[13] for two and a half months, which had prevented him from working or listening to classical music but had not interfered with other activities.

When the "operational shock" ended, David was able to work for about 3 hours a day, and he devoted this time to completing the manuscript of the book on Jamaican birds. His children were largely unaware of the seriousness of his illness until near the end. Two were going to school away from home, Paul at Bryanston and Andrew in his first year at the University of Aberdeen. Peter, in his second year at Oxford, was living in St. John's College, so only Catherine remained at home. The family was together for Christmas 1972, but other than the fact that their father was in obvious pain, which he sat with a hot water bottle to alleviate, the children were unaware that it was their last Christmas with him. They were surprised to learn from his Christmas letter that he had "a few weeks to a few years"[14] to live. David wanted to ensure that the children had a normal and happy Christmas, and he gave each of them LP albums. Catherine cherishes the two she received, Schubert's Octet and Britten's Spring Symphony, to this day.

According to Andrew Lack, his father's favorite gift was *The Faber Book of Religious Verse*. One is left to wonder whether Sir Walter Ralieigh's poem, *Epitaph* (possibly written the night before he was beheaded), which was reprinted in *The Faber Book of Religious Verse*, was of any consolation to David:

Even such is Time, which takes in trust
Our youth, our joys, and all we have,
And pays us with but age and dust;
Who in the dark and silent grave
When we have wandered all our ways,
Shuts up the story of our days:
And from which earth, and grave, and dust,
The Lord shall raise me up, I trust.[15]

FIGURE 20: David at Hatherton examining his favorite present, Christmas 1972. (Photograph courtesy of Lack family.)

In truth, it is difficult to know how David responded to his terminal illness. Theirs was a family that did not discuss emotions, and negative emotions were rarely expressed at home. There appeared to be a quiet resignation on his part, and Andrew said that his father did not fear death. At one point, in a reference to his favorite spectator sport, David declared that he had had "a good innings," but that hardly describes his emotional response to his impending death. His friend William Thorpe later recalled that during his illness David "was full of gratitude and happiness—never expressing for a moment either complaint or distress."[16]

Peter recalled that when his parents took him to Hatherton for his 21st birthday, 3 weeks after Christmas, it occurred to him that it might be the last birthday he celebrated with his father. Indeed, shortly after Peter's birthday David entered the Radcliffe Infirmary in Oxford, and he never returned home. His main priority remained the completion of the manuscript for the Jamaica book. He had an open and frank relationship with his consulting physician, and at one point David asked him whether he should work on improving his earlier drafts or on the "new bits." His doctor told him that he had best work on the new bits. One person who visited David in hospital remembers that he had to lie on his side to mitigate the intense pain, but that he continued to work on the book.

As Catherine recounted, "The last few weeks had a definite rhythm to them."[17] Elizabeth would drive Catherine and a neighborhood child to school each morning and then go directly to the hospital, where she would work with David on the manuscript of the Jamaica book. Someone else would drive Catherine home from school in the afternoon except on the one day a week that she went to the home of Niko and Lies Tinbergen for cello lessons with their daughter, Janet. Catherine remembered that her mother would often arrive early to pick her up from her lesson so that she could talk with Lies, "who was very helpful to her."[18] The Tinbergens and Lacks had been friends since Niko and Lies arrived in Oxford in 1949—a move that David had encouraged Niko to make following the war (see Chapter 5). The intensely private Elizabeth needed an old friend.

The end came quickly. Paul, who was at Bryanston, reported that his mother told him in a letter he received only 4 days before David's death that his father had less than 2 months to live. Elizabeth must have penned that letter at about the same time she called on Peter in St. John's College to tell him that she had just been advised that David had only a month or so to live. Peter recalled, "This I think was the only time I ever saw her break down. She was completely gutted as if it had finally dawned on her that she was going to lose him."[19] Trevor Williams, the Chaplain of Trinity College, served communion to David and Elizabeth late Sunday evening, March 11, and David slipped into a coma shortly thereafter. He died early on the morning of March 12 with Elizabeth still at his side.

The Rev. Gordon Owen, vicar of St. Peter's Church, Wootton, led a private crema-
tion service, with only Elizabeth present, 3 days after David's death. This was followed by a funeral service on March 17 at St. Peter's Church, again with Rev. Owen
officiating. The service was well attended, with many family members, friends, and
colleagues present. Peter Hartley, from David's early years in the EGI, came from
Suffolk, and Chris and Mary Perrins were present representing the contemporary
EGI. Niko and Lies also attended. St. Peter's, whose chancel dates from the early
14th century, is a charming and very small stone church with a steep, tiled roof and
bell turret. The churchyard in which David's ashes were interred after the service is
encircled by a stone fence. David's gravesite is marked with a small, flat stone on
which is inscribed, simply:

David Lambert

Lack

1910–1973

Elizabeth planted snowdrops and pale yellow primroses around the stone.

Elizabeth received 150 letters of condolence from friends and family members,
as well a few sympathy cards. Catherine commented, "These days it would be the
other way around, but she was much happier with people who had something to
say rather than just the sentiments."[20] One of the letters came from Leonard Elmhurst of Dartington Hall (see Chapter 2).

A memorial service led by Trevor Williams was held in the Trinity College chapel
on May 4, 1973, followed by a reception in the Senior Common Room. The final
hymn sung at the memorial service was David's favorite, *All People That on Earth do
Dwell*, verses 3 and 4 of which read as follows:

> O enter then His gates with praise;
> Approach with joy His courts unto;
> Praise, laud, and bless His Name always,
> For it is seemly so to do.
>
> For why? the Lord our God is good;
> His mercy is for ever sure;
> His truth at all times firmly stood,
> And shall from age to age endure.

Many friends, colleagues and family members attended the service, and most
stayed for the reception. The *Trinity College Record* reported, "He hoped that his

family and friends might wish to gather in Trinity for their own pleasure in meeting one another and not to mourn. Many came, and were glad to come, . . . but they missed him then, and miss him still."[21]

In his tribute to David in *Ibis*, longtime friend Ernst Mayr expressed that sense of loss well:

Considering his continued productivity it is obvious that David Lack's premature death is a great loss to science. Those who knew him personally feel, however, a much deeper loss. It is the loss of a unique human being, a person who was capable of total dedication to his work, to his family, and to the ideals in which he believed. In a world in which more and more people live only for the pleasures of each day and where complete loyalty to ideals and principles becomes rarer and rarer, one feels with particular acuteness the void caused by the death of someone whom one could truly admire and who stood for principles which give meaning to our existence.[22]

Either my interpretation of the small number of bird species on oceanic islands is wrong and, after all, they get there only rarely by chance, or we are extremely ignorant about the factors limiting the range of bird species. I suggest that the latter alternative is the more likely. . . . Perhaps therefore, the facts presented in this chapter chiefly illustrate our ignorance as to what limits the ranges of bird species. Even so, it seems surprising that so many as 12 of the 66 Jamaican land birds should be without close relatives or ecological counterparts on the two large adjoining islands only 150–180 km away.

—Island Biology as Illustrated by the Land Birds of Jamaica, pp 120–121

13

Island Biology

IN THE SPACE of less than 6 months in late 1972 and early 1973, two of the giants of 20th century ecology died, both felled prematurely by cancer. Robert H. MacArthur died of renal cancer on November 1, 1972, at the age of 42, and David Lack of non-Hodgkin's lymphoma on March 12, 1973, at the age of 62. Not only did they share the distinction of having made significant contributions in the field of ecology, but they also shared a close personal friendship which began when MacArthur spent a postdoctoral year at the Edward Grey Institute of Field Ornithology (EGI) in 1957–1958 (see Chapter 8). It is therefore somewhat ironic that a major book co-authored by MacArthur[1] rekindled Lack's interest in island biogeography and culminated in the writing of his last work, *Island Biology Illustrated by the Land Birds of Jamaica*, which was published posthumously by Blackwell Scientific Publications in 1976 as Volume 3 in their Studies in Ecology series.

Island Biology was the result of Lack's skepticism about some of the assumptions and conclusions in MacArthur and Wilson's book. After 25 years as Director of the EGI, he took his first sabbatical in Jamaica (see Chapter 12), in part to test some of his own ideas that differed from those put forward by MacArthur and Wilson. To appreciate the differences, it is necessary, first, to understand some of the empirical generalizations about island avifaunas (or island biotas generally), and second, to describe the theory proposed by MacArthur and Wilson to explain these observations.

Oceanic islands, those that have never been attached to a continent, are species depauperate; they have many fewer species than an equivalent area of the nearest continental land mass. Smaller islands have fewer species than larger islands located at the same distance from the mainland, and remote islands have fewer species than islands of similar size located closer to a continent. Finally, oceanic islands have a high proportion of endemic species, species that are found only on the island or in the archipelago to which the island belongs.

The theory proposed by MacArthur and Wilson attempted to explain these patterns using a simple graphical model describing the rate of colonization of an island by new species and the rate of extinction among species on the island as a function of the number of species already present. The model proposed that the colonization rate decreases with increasing numbers of species already present, whereas the extinction rate rises with an increasing number of species on the island. The point at which the lines representing these rates for a particular island intersect indicates an equilibrial species number for the island. Reasonable changes in the positions of the lines lead to changes in the equilibrial species number that match the empirical generalizations described earlier. For instance, a smaller near island would have a lower colonization rate than a larger near island because it represents a smaller target for colonizing individuals and a higher extinction rate than the larger island due to smaller population sizes (the likelihood of extinction is known to be inversely proportional to population size). As a result of these differences in rate, the intersection of the two lines predicts that the equilibrial species number on the smaller island will be lower than that on the larger island. The model implies that the species composition on a given island will show considerable change over time, with relatively infrequent colonization events being matched by an equal number of extinctions, thereby maintaining a dynamic equilibrium in species number.

As with some of his other books, David had already presented his own ideas in an earlier paper, this time the second Witherby Memorial Lecture at the annual meeting of the British Trust for Ornithology (BTO) in December 1969. The lectureship had been established the previous year by the BTO to honor Henry Forbes Witherby (1873–1943), a publisher and pioneering ornithologist. Witherby's company, H. F. and G. Witherby, was probably best known among ornithologists for its publication of *Handbook of British Birds* (1938–1941), an early field guide for which Witherby had also been the senior author. The firm had also published the first edition of Lack's *The Life of the Robin* in 1943 (see Chapter 2).

Lack began his lecture as follows:

St. Patrick explicitly banned snakes from Ireland, but the present state of the island suggests that his prejudices extended to woodpeckers, nightingales, and many other animals and plants. In all only three-fifths of the land and freshwater birds that breed regularly in Britain breed regularly in Ireland.[2]

He then summarized information about the land birds on many of the islands in the Atlantic Ocean, including Ireland, Iceland, the Canaries, the Azores, Puerto Rico, Jamaica, the Tristan group, and the islands of the Gulf of Guinea. He concluded that, at least for land birds, the problem of dispersal to islands was not a significant factor; rather, the difficulty of colonization on an island was the major factor responsible for limiting the number of species present. He attributed this difficulty to ecological factors, the most important of which was the relative ecological poverty of islands. Islands have fewer plant and insect species than continents and a lower structural diversity in the plant community. He suggested that this ecological poverty favors generalists over specialists and that early colonists to an island quickly adapt by broadening their ecological niches. The result is that a small suite of generalist bird species can exclude specialist species that reach the island later, thus precluding their successful colonization of the island. He concluded:

> I therefore consider that so far as land birds are concerned, islands are ecologically stable, which is contrary to the high rates of extinction and fresh colonisation . . . implied in the MacArthur-Wilson curves. I hold this view solely for primaeval habitats.[3]

Lack emphasized that the caveat expressed in the concluding sentence was important, because islands on which the natural habitats have been greatly altered by humans do tend to show high rates of species turnover.

In *Island Biology Illustrated by the Land Birds of Jamaica*, Lack used the data collected during his 10-month sabbatical visit to Jamaica, and his subsequent visit to Dominica (see Chapter 12), to bolster the case for his alternative to the MacArthur-Wilson model. The book was organized into two parts, the first part presenting an analysis and interpretation of his observations and the second describing the ecology and systematics of Jamaican land birds. He defined the land birds as those occupying terrestrial habitats, in contrast to those occupying freshwater or marine environments. His definition also excluded terrestrial-hunting hawks and included what he frequently referred to as "pigeons to passerines." The two parts of the book were followed by 29 appendices, in which much of the data was summarized.

Part One of *Island Biology* comprised 15 chapters and began with an introductory chapter in which Lack described his theory of the island biogeography of land birds, contrasting it with the MacArthur-Wilson model. Several ensuing chapters dealt in detail with the land avifauna of Jamaica and included descriptions of the island's principal natural habitats, the latter being identified broadly as semi-arid lowland forest, moist midlevel forest, and humid montane forest. Later chapters extended the analysis to other islands in the Greater Antilles chain (Cuba, Hispaniola, and Puerto Rico) and to the Lesser Antilles, particularly Dominica, where David spent 5 days in 1971 and he and Elizabeth spent 2 weeks in 1972 (see Chapter 12).

One of the reasons for choosing the island of Jamaica was that there was a reliable catalog of the avifauna of the island from the early 1800s, which permitted David to identify any changes in the composition of the land bird community that had occurred during the intervening 150 years. He reported that 68 land bird species (pigeons to passerines) occupied the island in the early 1800s, and that there had been two extinctions (a macaw and a nightjar) and no additions to the group since that time. Of the 66 remaining land bird species on the island, almost half occurred in all three main forest habitats, with only 5 species confined to a single forest habitat. Lack compared this observation with the situation in Honduras, the closest mainland country for which comparable data were available, and reported that, although Honduras had more habitat types than Jamaica, 64 percent of its 389 land bird species were confined to only one habitat. This remarkable contrast supported his assertion that island bird species have more generalized ecological niches than most mainland species.

Lack addressed the question of whether the small suite of generalized land bird species on Jamaica (66, versus 389 in Honduras) could be preventing more specialized species from colonizing the island. He posed the question, "Are all the available niches [on Jamaica] filled?" But he immediately acknowledged a major difficulty in attempting to answer that question: "There seems, however, to be no direct way of determining whether an ecological niche exists unless a bird is filling it."[4] This summarizes in a nutshell the persistent problem with the niche concept as an analytical tool, as opposed to its manifest utility as a heuristic device. Nevertheless, Lack found support for his contention that the ecological niches of Jamaica were basically filled.

One line of evidence concerned the geographical displacement of ecologically equivalent species among the islands. The most salient example presented by Lack involved the distributions of the nectar-feeding hummingbirds in the West Indies. On almost all of the mountainous islands in both the Greater and Lesser Antilles, there were three hummingbird species: a small species in both lowland and highland forests, a large species in lowland forest, and another large species in highland forest. On low-lying islands there were only two species, one small and one large. Exceptions to this general pattern included the fact that in the Greater Antilles, Cuba had only two species but Puerto Rico had five, whereas in the Lesser Antilles, two mountainous islands, Dominica and Martinique, had a fourth hummingbird species. Lack offered only ad hoc explanations for these exceptions. In another chapter, he identified many other examples of geographical displacement in other groups of land birds.

In a chapter comparing the land bird fauna of Jamaica with those of Cuba and Hispaniola, Lack again found that most of the Jamaican species occurred on one or both of the nearby islands, but for 12 of the 66 Jamaican species there was no ecological equivalent on the other islands. His conclusion regarding this fact serves as the epigraph to this chapter.

Lack based a second line of support for the proposition that there were no vacant ecological niches on Jamaica and other West Indian islands on the evidence that dispersal of birds to Jamaica and the other islands by mainland species occurred quite frequently. Despite the relative paucity of observers on the islands, reports of such potential colonists were numerous, and Lack related that he had observed such dispersers himself during his relatively short stays. He claimed that the frequency of such dispersal without attendant colonization supported his contention that the ecological niches of the islands were essentially filled.

A chapter on the migrant wood warblers from North America that winter on Jamaica raised a question that Lack did not adequately answer. In the chapter dealing with Jamaican warblers, he described how the 18 wintering warbler species were ecologically isolated from each other in their foraging niches. However, he did not address the question of what happened to those niches during the Jamaican summer, when the warblers were breeding in North America.

In a quite critical review in *Auk*, Robert Ricklefs acknowledged the value of *Island Biology*, because it presented an alternative theory to the prevailing MacArthur-Wilson model, but chided the author because the book was "overly long for its accomplishments."[5] He identified the principal differences between Lack's theory and that of MacArthur and Wilson but accused Lack of emphasizing evidence that favored his theory while ignoring or explaining away countervailing evidence. He concluded, "I was saddened to find that *Island Biology* is frequently little more than an attempt to marshall [*sic*] weak evidence behind an outdated concept of island biogeography."[6]

The MacArthur and Wilson model of island biogeography continues to enjoy wide acceptance among ecologists, and it has had a major impact on conservation practices. Nature preserves are often "habitat islands" surrounded by an "ocean" of human-altered habitats. The theory has helped to inform conservationists about how to best allocate their resources to maximize the preservation of species. Lack's alternative hypothesis and his "close" approach to understanding the species composition on islands still have adherents, however.

The balance of this chapter summarizes some of Lack's most significant scientific contributions and identifies some of the most important formative influences in his development as a scientist. I will also attempt to answer the question, "Who was David Lack?"

MAJOR INFLUENCES

Although it is difficult to identify with confidence the major influences that shape the development of a great scientist, several contributors to Lack's success as a scientist were clearly significant. Four of the most important individuals–Julian Huxley, Ernst Mayr, Reg Moreau, and Niko Tinbergen–were profiled in Chapter 11, along with their major influences on Lack. Two other influences were institutions, and one was a person whom he may never have met.

Lack himself identified one of his earliest influences: "I . . . learned to think in an evolutionary way about birds from the semi-popular books of W. P. Pycraft."[7] William Plane Pycraft, who was born in Great Yarmouth, Norfolk, in 1868, was a self-taught naturalist who worked in several museums before concluding his career at the Natural History Museum in London. Two of Pycraft's "semi-popular" books were *The Story of Bird-Life* (George Newes, 1900) and *A History of Birds* (Methuen, 1910). Pycraft died in 1942, probably without ever meeting the man whose evolutionary perspective he helped to shape. Lack had read Pycraft's books by the time he was a schoolboy at Gresham's, and his evolutionary perspective, although somewhat naïve, was evident in his paper that won the school's Holland-Martin Natural History Prize in 1926 (see Chapter 1).

The two institutions that had profound influences on Lack's development were Gresham's School and Dartington Hall School. Today Gresham's School is a coeducational public school that serves primarily students from the Norfolk region. A 2010 ranking of U.K. high schools based on A-level results listed Gresham's as no. 416 of 1000 ranked schools.[8] Although this placement hardly suggests greatness, the Gresham's of Lack's day was different. The list of students whose attendance was contemporaneous with David's is truly remarkable (see Chapter 1). More to the point, however, is the fact that two of the 20th century's pioneering ecologists attended Gresham's within a few years of each other: G. Evelyn Hutchinson (1917–1921) and David Lack (1924–1929). As David would have said, this could hardly have been due to coincidence. What was it about the Gresham's School of the 1920s that helped it produce two such eminent ecologists?

Much of the answer to that question lies with the vision of George W. S. Howson, the progressive headmaster of Gresham's from 1900 until his death in 1919. Howson elected to deemphasize the classics, the core of the traditional public school curriculum of the day, and instead placed more emphasis on the sciences and the arts. He also chose to greatly reduce the status of competitive sports, another staple of English public schools. Instead, he encouraged the development of clubs and societies that reflected students' interests. The largest of these was the Natural History Society, to which more than two-thirds of the boys belonged during Lack's time at Gresham's.

The Natural History Society had several sections covering both the physical and biological sciences, and it published an annual report that provided opportunities for boys to publish some of their findings. In addition, the society awarded an annual Holland-Martin Prize for the best student research paper. Both Hutchinson and Lack won the Holland-Martin Prize. By the time David left Gresham's for Cambridge, he had decided to become a zoologist rather than follow his father's footsteps into medicine. J. R. Eccles, deputy headmaster and Howson's protégé, became headmaster after Howson's death and remained in that post throughout Lack's time at Gresham's.

The impact on Lack of Dartington Hall School, and indeed of the Dartington Hall community generally, was quite different. Dartington Hall was an egalitarian and self-consciously Utopian community. All members of the community, from farmhands to the owners, Leonard and Dorothy Elmhurst, were encouraged to attend the regular Sunday Night Meeting, which featured speakers such as George Bernard Shaw, Julian Huxley, Bertrand Russell, and many other notable writers and intellectuals. All the children of the community were educated in the Dartington Hall School, which was coeducational from its inception, alongside the children of many of the liberal intelligentsia of the day (see Chapter 2). The community also had a strong commitment to the arts, particularly the performing arts.

David clearly enjoyed his participation in the Dartington Hall community during the 7 years he spent there. His interests in literature and music were nourished, and although he was quite shy, he was embraced by the members of the community. He also displayed an interest in political events and was an active member of a pacifist group at Dartington Hall. Although his interests in literature and music never waned, he showed virtually no interest in politics after World War II.

The most lasting influences of Dartington Hall, however, were associated with his teaching. Some of Lack's colleagues and former students remarked that he remained "schoolmasterly" the rest of his life, usually implying that he was needlessly didactic and somewhat condescending. A more positive interpretation of the lessons in communication that David learned at Dartington Hall is that he was striving to impart his scientific insights to a frequently indifferent and sometimes hostile audience. He was clearly attempting to communicate his science to a wide audience in books such as *The Life of the Robin, Swifts in a Tower, Enjoying Ornithology*, and *Evolution Illustrated by Waterfowl*.

The progressive headmaster of Dartington Hall School, Bill Curry, believed strongly in experiential education. This philosophy provided Lack not only the opportunity but also the encouragement to design research projects that would engage students. Clearly, the most significant of these projects was his 4-year study of the robin. Dartington Hall also granted Lack a sabbatical leave that enabled him to travel to the Galapagos Islands and to spend time at the California Academy of Sciences.

MAJOR SCIENTIFIC CONTRIBUTIONS

David Lack's contributions to science were numerous and included both his own scientific achievements and his institutional impact, principally on the development of the EGI as its Director from 1945 until 1973. He also had an enormous indirect impact through the myriad achievements in ornithology and ecology of his D.Phil. students, both in Great Britain and abroad.

Lack's most important theoretical contributions, the ones on which his recognition as the father of evolutionary ecology rests, were contained in his two highly significant 1947 publications (see Chapter 3). In *Darwin's Finches*, he identified the need for ecological segregation in the process of speciation in birds. In his paper on clutch size, he proposed that natural selection would always act to maximize the reproductive output of the individual, a departure from the widely held view that the reproductive rate in a species was selected to balance mortality. His theory of clutch size determination in birds represented the first optimization model in ecology.

Another novel contribution that has generated sustained interest, essentially a corollary of his clutch size hypothesis, was the brood reduction hypothesis, proposed to explain hatching asynchrony in birds. His most direct formulation of that hypothesis came in his 1954 book, *The Natural Regulation of Animal Numbers* (see Chapter 5). In addition, his conclusion, developed in *The Natural Regulation of Animal Numbers*, that animal populations are regulated by density-dependent mechanisms, primarily competition for food, had an enormous impact on population ecology. Although he owed the idea of density-dependent population regulation to A. J. Nicholson, his carefully reasoned argument that food was usually the limiting factor led to much controversy and additional research.

The Natural Regulation of Animal Numbers also represented a second facet of Lack's abilities, his talent at amassing and coherently interpreting a large body of disparate information–his ability to synthesize. Three of his later books, *Population*

Studies of Birds, Ecological Adaptations for Breeding in Birds, and *Ecological Isolation in Birds*, demonstrated this ability to synthesize, and all received high praise, particularly *Ecological Adaptations for Breeding in Birds*.

Lack's most important empirical contributions probably derived from his multifaceted study of migration. Beginning with his wartime discovery, along with George Varley, that flying birds were detected by radar (see Chapter 2) and with his observation in the Pyrenees, along with his new bride Elizabeth, of high mountain migration of birds and insects (see Chapter 6), he made numerous contributions to the understanding of migration patterns by both direct visual and radar-aided observation. Speaking at the Lack Centenary Symposium in Oxford in July 2010, Thomas Alerstam identified several areas in which Lack made significant contributions: height of migration, migrational drift, migratory orientation, weather migration, and migratory patterns throughout the year.[9]

His studies of the robin and the swift also resulted in important observations. In his study of the robin, he was among the first to document the high rates of both juvenile and adult mortality in small birds, as well as sex and age differences in migration in the species (see Chapter 2). His studies of the swift identified several important adaptations for life in the air, perhaps most saliently the torpor and suspended development of nestlings during prolonged periods of inclement weather when the adults were unable to forage for flying insects.

When David Lack assumed directorship of the EGI in 1945, the institute had a permanent staff of one—its retiring Director, W. B. Alexander. With increased funding from Oxford University and the encouragement of Alister Hardy, Head of the Department of Zoology, David quickly developed a new vision for the institute: "With the ending of the war, the Institute is re-commencing active field work on pure research, primarily on bird ecology. The main center for this work will be on the Wytham estate."[10] Under David's almost 28 years of leadership, the EGI was transformed from obscurity into an internationally recognized center of research in avian biology. That reputation has been extended and enhanced by David's successors.

At the time of David's death, Oxford University was experiencing one of its periods of economic stringency, and there was a question about whether to appoint another director of the EGI. In 1974, however, Chris Perrins, one of David's D.Phil. students who had remained at the EGI since completing his doctorate, was appointed to the post, and under his very capable leadership the EGI continued to flourish. By the time Perrins retired in 2002, he had supervised more than 90 D.Phil. students in ornithology, from countries throughout the world. Upon Perrins' retirement, Ben Sheldon was chosen to head the institute, and after the creation of the endowed Luc Hoffmann Chair in Field Ornithology, he was also appointed first holder of the Luc Hoffmann Professorship. Thanks to the energetic and creative leadership of Perrins and Sheldon, the EGI continues to flourish as an international center of research in avian biology.

The long-term study of tits in Wytham Woods inaugurated by Lack in 1947 represents another of his major scientific contributions. The study, which has been expanded several times since its inception, continues to the present day, and has resulted in hundreds of scientific papers and numerous D.Phil. theses. Lack chose the tits for long-term study after visiting the Netherlands in February 1946 and observing the long-term tit study there (see Chapter 5). He had a firm belief in the value of long-term studies no doubt in part based on what Thomas Alerstam referred to as Lack's law: the tendency in an extended series of observations for the first set to be abnormal, thus biasing the observer against the true explanation.[11] In Lack's mind, the principal antidote to the bias caused by these perverse vagaries of nature was the long-term study.

Another of Lack's first initiatives after he became Director of the EGI was the creation of the annual Student Conferences in Bird Biology, the first of which was held in December 1946 (see Chapter 5). The conferences have been held annually since their inception, usually during the first week in January. Their basic structure has remained essentially as Lack devised it. The primary purpose is to provide graduate and undergraduate students the opportunity to read research papers, with three or four senior scientists being invited to present plenary talks, usually on the central theme of the conference. There were 37 participants in the first conference in 1946, all from Great Britain; recent conferences have hosted more than twice that number and have included many students from European nations as well as a few from the Far East. Many of Britain's preeminent ornithologists of the past half-century gave their first presentation at a Student Conference. David and his successor Directors also used the conferences to scout for promising students. Several of David's D.Phil. students told me that he first approached them about coming to the EGI to pursue the doctorate after they had presented their research at a Student Conference.

Finally, a significant element of Lack's influence resides in the achievements of the doctoral students he supervised (see Chapter 10). Although most remained in Great Britain, others spent all or part of their careers in Australia, Canada, Denmark, Ecuador (the Galapagos), Iran, Israel, New Zealand, and the United States. Most also continued to work primarily on birds, but several made significant contributions in other related fields, particularly ecology, animal behavior, and conservation biology. Some indication of the significance of their contributions is provided by the major honors and awards they have received. Three (Hinde, Newton, and Perrins) have been elected Fellows of the Royal Society, and Hinde is also an Honorary Fellow of the British Academy, Honorary Foreign Associate of the National

Academy of Sciences (USA), and Foreign Honorary Member of the American Academy of Arts and Sciences. Four (Evans, Newton, Perrins, and Snow) received the Godman-Salvin Medal from the British Ornithologists' Union (BOU), and two (Newton and Stonehouse) received the BOU's Union Medal.

HONORS AND AWARDS

Ironically, the first award received by the thoroughgoing Englishman, David Lack, was bestowed by the American Ornithologists' Union. He was awarded the William Brewster Medal in 1948, which is presented annually to the author (or co-authors) of an exceptional body of work on birds of the Western Hemisphere. The medal was given in recognition of Lack's studies of the Galapagos finches (see Chapter 3). One of Lack's D.Phil. students, David Snow, shared the Brewster Medal with his wife Barbara in 1972.

Lack was elected a Fellow of the Royal Society in 1951. Again the award was given primarily in recognition of his Galapagos Islands work:

> Dr. David Lack: . . . distinguished for his studies on the behaviour and evolution of birds, especially the formation of species and races of the finches of the Galapagos Islands.[12]

David commented later about this honor, "I suppose I may be one of the last to be elected for research done as an amateur."[13]

At the Centennial Meeting of the BOU in 1959, David was awarded the God-man-Salvin Medal. This medal, the highest recognition bestowed by the BOU, has been awarded periodically since 1922 as a "signal honor for distinguished ornithological work." There have been 27 recipients of the award in the 90 years of its existence.

Probably the crowning formal recognition of Lack's scientific contributions was his receipt of the Darwin Medal. The Darwin Medal is awarded biennially by the Royal Society of London "for work of acknowledged distinction in evolution, population biology, organismal biology and biological diversity." First given in 1882, the medal was created in memory of Charles Darwin; it bears a profile image of the great evolutionary biologist on one side and the inscription "Carolus Darwin" with his dates (MDCCCIX and MDCCCLXXXI) on the other. David was awarded the Darwin Medal at the Anniversary Meeting of the Society held in London on November 30, 1972. Sir Alan Hodgkin, who had first met his fellow Gresham's student on the salt marshes of Cley, Norfolk, one Sunday afternoon in the fall of 1927 (see Chapter 1), was then the President of the Royal Society and presented the award. Unfortunately, David was too ill to attend the ceremony, and Hodgkin, who wanted to read the full citation accompanying the medal but could not do so unless someone was present to collect the award, urged him to send someone in his place. Peter Lack, David's oldest child, attended and accepted the Darwin Medal for his father. David's brother Christofer and his wife also attended the ceremony. The citation was read by Hodgkin:

The DARWIN MEDAL is awarded to Dr. D. Lack, F.R.S.

The researches of David Lack, Director of the Edward Grey Institute of Field Ornithology, Oxford, have greatly advanced our understanding of ecological speciation in birds, and of the breeding habits and other factors that limit their numbers in natural populations and determine the separation of closely related species. His gift for keen field observation and original interpretation were early shown in his *Darwin's finches* (1947), which did much to promote the interaction of ecological and evolutionary studies, and which still provides a continuing stimulus to new generations of readers.

In *The natural regulation of animal numbers* (1954, reprinted 1970), he was able, by careful weighing of evidence, to set out arguments for the view that density-dependent processes are the prime regulators of natural populations. Twelve years later, in 1966, he developed his arguments further in *Population studies of birds*, a sequel based on new researches by a number of workers who had been in large measure stimulated by his example. These and his other books, which include *Ecological adaptations for breeding in birds* (1968) and *Ecological isolation in birds* (1971), are models of what can be achieved in organizing complex data by a writer with a clear and logical mind, aided by a graceful and stimulating mode of presentation. He has successfully advanced in them an intention well set out in his own words: to avoid intricate and nebulous theory, and also to avoid an arid enumeration of bare facts.

Working with such distinction in a tradition of which Charles Darwin remains the prime exemplar, David Lack is pre-eminently fitted for the award of the Darwin Medal.[14]

Other important recognitions of Lack's scientific achievements included his election as President of the XIV International Ornithological Congress, held in Oxford in 1966 (see Chapter 8), and his election as President of the British Ecological Society, 1964–1965.

WHO WAS DAVID LACK?

A significant handicap for a biographer attempting to chronicle the life of his subject is not having known the person directly. I was never fortunate enough to meet
David Lack, although his work had a major impact on my development as an avian
ecologist. As a result, I am forced to try to put together a picture of the man from
personal interviews with colleagues, former students, and family members. One
question I asked virtually everyone I interviewed was, "If you could use three descriptors to characterize David Lack, what would they be?" The answers to that
question, as well as the interviewees' recollections of their interactions with David,
provide the basis for this section.

A number of people remembered David as socially awkward, tending to relate to
other people primarily through his work rather than on a personal level. Descriptors
included "shy," "aloof," "not terribly social," "very modest," "eccentric," and "prickly."
At tea time in the EGI, the topics he introduced for discussion usually related to
whatever he was working on at the time. Tea-time discussions provided him with a
great opportunity to hone his own ideas and to generate new avenues for exploration. One person recalled that he was "great at picking other people's brains"; little
wonder that Reg Moreau's sense of humor helped to lighten the atmosphere. One
person who worked with David over a long period believed that there had been a
change in his behavior during his long tenure as Director of the EGI. He described
David as being "abrasive" and "dismissive" and "not very tolerant of people who got
it wrong" during his first few years at the Institute, but as his children grew older,
he became "reclusive," focused on his work and on his family. One of Lack's later
D.Phil. students recalled that he sometimes began tea-time discussions with
something about one of his children.

Several people mentioned his intellectual capacity, with descriptors ranging from
"intellectually sharp" to "genius." One D.Phil. student called him "brilliant," but another, Bernard Stonehouse, said he was not brilliant. When I questioned Stonehouse
about this characterization, he said, "I don't think he thought of himself as brilliant.
In fact, he distrusted brilliant people."[15] It is possible that this self-perception derived
from Lack's first-hand experiences with people such as Bertrand Russell, George
Bernard Shaw, H. G. Wells, and Julian and Aldous Huxley during his time at Dartington Hall. Perhaps he recognized that he did not have the dazzling verbal facility
of such men, and instead understood that his strength lay in his methodical, logical
approach to problems. In this regard, he was probably more like the plodding
Charles Darwin than the brilliant T. H. Huxley.

David's leadership style in the EGI was decidedly hierarchical, perhaps a carryover
from his days as a schoolmaster. His staff and students were not permitted to address

him by his given name, but instead were required call him "Dr. Lack." Only some time after a student had completed his doctorate was he permitted to address the Director as "David." Lack often quoted his friend, V. C. Wynne-Edwards, who also required those working under him to address him as "Professor." Wynne, as he was called by close friends, had the following dictum: "I don't allow those working for me to call me by my given name. I might have to sack one of them, and I would find it difficult to sack someone who uses my given name."[16]

One characteristic that was repeatedly identified was David's intense focus on whatever task he was working on. One person said that he had "ferocious powers of concentration." This focus was reflected in the fact that the tea-time topics he introduced for discussion were usually directly related to his current project.

David had little interest in his personal appearance. He often wore a long, somewhat shabby raincoat, and his shoes were usually scuffed.

One consistent interest, recognized by virtually all his students, colleagues, and family members, was his love of birds–"Mad keen on birds," as Peggy Varley expressed it (see Chapter 8). One of his children said that he was more interested in birds than in people. Birds were his true passion from the time he was a schoolboy.

Two of David's doctoral students during the late 1960s were avid birders and spent many weekends looking for birds in the Oxford area. They both related that David would often approach them somewhat sheepishly on Monday mornings, inquiring about what good birds they had seen that weekend under the pretext of wanting to relay the information to Peter, whose interest in birds was intensifying. One Monday morning, David was told by Derek Scott that he had seen a whiskered tern at a local gravel pit, after which David disappeared, as usual, into his office and began hammering away on his typewriter. Before long, however, the hammering stopped and David emerged from his office, saying, "It's such a fine day, and I promised Elizabeth a drive in the countryside. I believe I'll take her today." "So off he went," said Derek. "He saw the bird and thanked me profusely when he returned—happy as a schoolboy."[17]

David Lambert

Lack

1910–1973

NOTES

CHAPTER 1

1. K. Johnson, *The Ibis:* Transformations in a twentieth century British natural history journal, *Journal of the History of Biology* 37:515–555 (2004).

2. E. Lack, Interview by Ted Anderson, October 16, 2007.

3. A. Lack, Personal communication, September 14, 2009.

4. D. Lack, My life as an amateur ornithologist, *Ibis* 115:422 (1973).

5. A. Lack, *loc. cit.*

6. A. Lack, Personal communication, April 20, 2009.

7. D. Lack, *loc. cit.*

8. W. H. Auden, The liberal fascist, in E. Mendelson (ed.), *The English Auden: Poems, Essays and Dramatic Writings 1927–1939* (London, Faber and Faber, 1977), pp 322–323.

9. H. Carpenter, *Benjamin Britten: A Biography* (New York, Charles Scribner's Sons, 1992), p 27.

10. D. Lack, *loc. cit.*

11. Ibid., p 423.

12. Ibid.

13. Ibid.

14. A. Hadgkin, In appreciation, *Ibis* 115:431 (1973).

15. D. Lack, Three birds of Kelling Heath: Being the life-history of the nightjar, the redshank, the ringed plover, Entry for the Holland Martin Natural History Prize, 1926. David Lack archive, Alexander Library of Ornithology, Oxford (hereafter referred to as Lack archive), folder 5.

16. D. Lack, Bird diary, volume I, Lack archive, folder 8.

17. D. Lack, *My life as an amateur ornithologist, op. cit.*, p 424.

18. Ibid.

19. J. S. Gardner, Letter to Tutor at Magdalene, June 16, 1933, Magdalene College, Cambridge, archive.

20. D. Lack, Letter to Tony Diamond, June 10, 1966, Courtesy of A. W. Diamond.

21. J. H. Bell, quoted in T. E. B. Howarth, *Cambridge Between Two Wars* (London, Collins, 1978), p 156.

22. D. Lack, Bear Island in 1932, *Magdalene College Magazine* No. 70, 10(3):79–81 (1933).

23. D. Lack and B. B. Roberts, Notes on Icelandic birds, including a visit to Grìmsey, *Ibis* 4:801 (1934).

24. J. B. Charcot, Letter to B. B. Roberts, C. Bertram, and D. Lack, Undated, Lack archive, folder 23.

CHAPTER 2

1. P. Thomas, Personal communication, March 17, 2011.

2. H. N. Southern, The population dynamics of birds, *Journal of Animal Ecology* 13:86 (1944).

3. V. Bonham-Carter, *Dartington Hall: The History of an Experiment* (Ithaca, N.Y.: Cornell University Press, 1958), p 21.

4. M. Young, *The Elmhursts of Dartington: The Creation of an Utopian Community* (London, Routledge & Kegan Paul, 1982), p 136.

5. Ibid., p 131.

6. Ibid., p 134.

7. L. K. Elmhurst, Forward to "The school," in V. Bonham-Carter, *op. cit.*, p 160.

8. W. B. Curry, "The school," in V. Bonham-Carter, *op. cit.*, p 198.

9. Ibid., p 202.

10. D. Lack, Letter to L. Elmhurst, Dartington Hall Archive, Courtesy of Peter Lack.

11. Ibid.

12. R. Leacock, Manuscript of Memoir, Courtesy of R. Leacock.

13. E. Ibbotson, My best teacher, *Times Education Supplement*, March 3, 2006.

14. A. H. Wolff, Personal communication, January 1, 2011.

15. D. Lack, Obituary of R. E. Moreau, *Ibis* 112:557 (1970).

16. D. Lack, Trip to Tanganyika 1934 (typescript), p 5, Lack archive, folder 27.

17. Ibid., p 4.

18. Ibid., p 15.

19. Ibid., p 16.

20. Ibid.

21. D. Lack, Obituary of R. E. Moreau, *loc. cit.*

22. D. Lack, Trip to Tanganyika 1934, *op. cit*, p 17.

23. Ibid., p 25.

24. Ibid., p 23.

25. Ibid.

26. Ibid., p 30.

27. Ibid., p 44.

28. Ibid., p 45.

29. Ibid., p 46.

30. Ibid., p 47.

31. Ibid., p 48.

32. Ibid., p 50.

33. Ibid., p 52.

34. Ibid., p 55.

35. Ibid.

36. Ibid, pp 57–58.

37. Ibid., p 60.

38. Lack usually identified people in his journals by the first initial of their given name. Despite many attempts, I have not been able to identify "P."

39. D. Lack, Trip to California 1935 (typescript), p 8, Lack archive, folder 27.

40. Ibid., p 9.

41. Ibid., p 18.

42. Ibid., p 23.

43. Ibid., p 24.

44. Ibid., p 29.

45. Ibid., pp 31–32.

46. Ibid., pp 34–35.

47. E. Kabraji, "Dartington Hall School in wartime," Unpublished manuscript, Courtesy of Etain Kabraji Todds.

48. D. Lack, *Ode to the Deare Departed*, Unpublished poem, in E. Kabraji, Ibid.

49. D. Lack, My life as an amateur ornithologist, *Ibis* 115:428 (1973).

50. Ibid.

51. D. Lack, *Radar Echoes from Birds*, Army Operational Research Group Report No. 257, February 1945, Courtesy of T. Alerstam.

CHAPTER 3

1. D. Lack and L. Lack, Territory revisited, *British Birds* 27:179–199 (1933).

2. P. R. Lowe, The finches of the Galapagos in relation to Darwin's conception of species, *Ibis* 78:310–321 (1936).

3. D. Lack, My life as an amateur ornithologist, *Ibis* 115:427 (1973).

4. D. Lack, Handwritten journal of the Galapagos expedition, Lack archive, folder 36.

5. Ibid.

6. Ibid.

7. Ibid.

8. Ibid.

9. Ibid.

10. Ibid.

11. Ibid.

12. J. Andermeyer, *My Father's Island: A Galapagos Quest* (New York, Viking, 1989).

13. D. Lack, Variation in the introduced English sparrow, *Condor* 42:239–241 (1940).

14. D. Lack, *The Galapagos Finches (Geospizinae), A Study in Variation*, Occasional Papers of the California Academy of Sciences 21:1–159 (1945), p 116.

15. Ibid., p 117.

16. Ibid.

17. Ibid., p 119.

18. Ibid., p 120.

19. D. Lack, *Darwin's Finches*, (Cambridge, Cambridge University Press, 1947), p 147.

20. D. Lack, My life as an amateur ornithologist, *op. cit.*, p 429.

21. Ibid., p 430.

22. O. S. Pettingill, Jr., *A Laboratory and Field Manual of Ornithology* (Minneapolis, Burgess, 1946).

23. D. Lack, *The Natural Regulation of Animal Numbers* (Oxford, Clarendon Press, 1954), p 32.

CHAPTER 4

1. D. Lack, *Robin Redbreast* (Oxford, Clarendon Press, 1950), p 5.

2. Ibid., p 34.

3. Ibid., p 142.

4. D. Lack, The effect of the exodus from St. Kilda upon the island's fauna and flora: Interesting changes observed during a recent visit, *The Illustrated London News*, December 26, 1931, p 1054.

5. Ibid.

6. D. Lack, Of birds and men, *New Scientist* 41:121–122 (1969), p 122.

7. D. Lack, Mr. Lawson of Charles. *American Scientist* 51:12–13 (1963), p 13.

8. D. L. Lack, Population, biological, *Encyclopedia Britannica* 14th ed. (1980).

9. A. Lack, Personal communication.

CHAPTER 5

1. I. Newton, Population regulation in birds: Is there anything new since David Lack?, *Avian Science* 3:75–84 (2003).

2. D. Lack, *The Natural Regulation of Animal Numbers* (Oxford, Clarendon Press, 1954), p 5.

3. Ibid.

4. Ibid., pp 40–41.

5. D. Lack and E. Lack, The breeding biology of the swift *Apus apus*, *Ibis* 93:501–546 (1951).

6. D. Lack, *op. cit.*, p 212.

7. Ibid., p 223.

8. J. B. S. Haldane, Special review, *Ibis* 97:375 (1955).

9. Ibid., p 376.

10. V. C. Wynne-Edwards, The dynamics of animal populations, *Discovery*, October 1955, p 434.

11. Anonymous, "The proposed Institute of Field and Economic Ornithology at Oxford" (typescript), EGI archive, p 1.

12. Ibid., p 3.

13. Letter from Secretary, O. U. Committee for Ornithology, July 4, 1945, History of Alexander Library archive, Alexander Library of Ornithology, Oxford.

14. H. N. Southern, Letter to W. B. Alexander, July 9, 1945, History of Alexander Library archive, Alexander Library of Ornithology, Oxford.

15. B. W. Tucker, Telegram to W. B. Alexander, July 27, 1945, History of Alexander Library archive, Alexander Library of Ornithology, Oxford.

16. P. Crowcroft, *Elton's Ecologists: A History of the Bureau of Animal Population* (Chicago: University of Chicago Press, 1991), p 44.

17. N. Tinbergen, Letter to D. Lack, November 2, 1945, Lack archive, folder 403.

18. J. Krebs, Preface, in P. S. Savill, C. M. Perrins, K. J. Kirby, and N. Fisher (eds.), *Wytham Woods: Oxford's Ecological Laboratory* (Oxford: Oxford University Press, 2010).

19. N. Tinbergen, Letter to D. Lack, January 30, 1946, Lack archive, folder 403.

20. Ibid.

21. R. Moreau, Autobiography, *Ibis* 112:5505(1970).

22. I. Goodbody, Personal communication, July 21, 2012.

23. G. L. Jepsen, Letter to D. Lack, July 13, 1946, Lack archive, folder 391.

24. E. Mayr and W. B. Provine (eds.), *The Evolutionary Synthesis: Perspectives on the Unification of Biology* (Cambridge, Mass.: Harvard University Press, 1980), p 1.

25. G. L. Jepsen, E. Mayr, and G. G. Simpson (eds.), *Genetics, Paleontology, and Evolution* (Princeton, N.J.: Princeton University Press, 1949).

26. P. Griffin, *St. Hugh's: One Hundred Years of Women's Education in Oxford* (Basingstoke, Hampshire, U.K.: Palgrave Macmillan, 1986), p 126.

27. D. W. Snow, Unpublished manuscript, Courtesy of D. W. Snow.

28. D. Lack, EGI Annual Report, 1947, p 9.

29. J. Morris, *Oxford* (Oxford: Oxford University Press, 1990), p 78.

30. A. C. Hardy, Report of the Linacre Professor as Head of the Department of Zoological Field Studies for the year ended July 31, 1948, p 3.

31. J. K. A., Obituary, *Ibis* 108:289 (1966).

CHAPTER 6

1. D. Lack, *Swifts in a Tower* (London: Methuen, 1956), pp 16–17.

2. Ibid., p 56.

3. Ibid., p 208.

4. Ibid.

5. Ibid., p 213.

6. E. Lack, Interview by Ted Anderson, October 16, 2007.

7. D. Lack, *Enjoying Ornithology* (London: Methuen, 1965), p 31.

8. C. Lack, Personal communication, September 20, 2010.

9. D. Nichols, Personal communication, December 10, 2011.

10. G. Lambrick, Interview by Ted Anderson, April 26, 2008.

11. Paul Lack, Interview by Ted Anderson, October 18, 2009.

12. Ibid.

13. A. Lack, Personal communication, April 19, 2010.

14. A. Lack, Personal communication, September 1, 2012.

15. E. Lack, Interview by Ted Anderson, October 16, 2007.

16. A. Lack, Personal communication, September 1, 2012.

17. Peter Lack, Interview by Ted Anderson, April 28, 2008.

18. A. Lack, Personal communication, June 1, 2010.

19. Ibid.
20. Ibid.
21. Paul Lack, *loc. cit.*
22. D. Nichols, *loc. cit.*

CHAPTER 7

1. John Wren-Lewis, Letter to David Lack, August 20, 1953, Lack archive, folder 380.
2. D. Lack, Manuscript of "Evolution and Christian Belief," Lack archive, folder 379.
3. Peter Medawar, Letter to David Lack, January 18, 1954, Lack archive, folder 380.
4. David Lack, Letter to editors at Methuen, September 18, 1956, Lack archive, folder 382.
5. Ibid.
6. D. Lack, *Evolutionary Theory and Christian Belief, the Unresolved Conflict* (London: Methuen, 1957), p 12.
7. Ibid.
8. Ibid., p 114.
9. Ibid., p 116.
10. D. Lack, *Evolutionary Theory and Christian Belief, the Unresolved Conflict*, 2nd ed. (London: Methuen, 1961), p 129.
11. D. Lack, Manuscript of "Natural Selection and Human Nature," Lack archive, folder 384.
12. D. Lack, T. H. Huxley and the nature of man, in *Enjoying Ornithology* (London: Methuen, 1965), p 218.
13. D. Lack, The conflict between evolutionary theory and Christian belief, *Nature* 187:98–100 (1960).
14. E. Lack, Interview by Ted Anderson, October 16, 2007.
15. A. Lack, Personal communication, May 6, 2009.
16. D. Lack, Manuscript of "Evolution and Christian Belief," Lack archive, folder 379.
17. R. Darwall-Smith, Personal communication, March 11, 2010.
18. A. Lack, Personal communication. August 27, 2009.
19. D. Lack, Letter to Michael Wilson, March 1958, Lack archive, folder 383.
20. P. Hartley, Letter to David Lack, January 18, 1956, Lack archive, folder 381.
21. D. and E. Lack, "Some letters on the family at church," Unpublished manuscript, Lack archive, folder 381.

CHAPTER 8

1. M. E. Varley, Interview by Ted Anderson, April 23, 2008.
2. A. Lack, Personal communication. April 13, 2010.
3. Paul Lack, Interview by Ted Anderson, October 18, 2009.
4. D. Lack, *Enjoying Ornithology* (London: Methuen, 1965), p 238.
5. Ibid., p 239.
6. Ibid., p 241.
7. D. Lack, EGI Annual Report, *Bird Study* 9:136 (1962).
8. K. Johnson, *The Ibis:* Transformations in a twentieth century British natural history journal, *Journal of the History of Biology* 37:515–555 (2004).
9. D. Lack, EGI Annual Report, *Bird Study* 10:140 (1963).

10. C. G. Phillips, David Lack, Sc.D., F.R.S. (obituary), *Trinity College Record 1972–1973*, p 11.

11. D. Lack, Presidential address, *Proceedings of the XIV International Ornithological Congress* (1967), p 3.

12. O. L. Austin Jr., Special review, *Auk* 84:145 (1967).

CHAPTER 9

1. J. B. Hagen, *An Entangled Bank: The Origins of Ecosystem Ecology* (New Brunswick, N.J.: Rutgers University Press, 1992.

2. H. G. Andrewartha and L. C. Birch, *The Distribution and Abundance of Animals* (Chicago: University of Chicago Press, 1954).

3. D. Lack, *Population Studies of Birds* (Oxford: Clarendon Press, 1966), p 2.

4. D. Chitty, What regulates bird populations? *Ecology* 48:698–701, p 698.

5. D. Chitty, Handwritten note on reprint of "What regulates bird populations?" sent to D. Lack, Lack archive, folder 222.

6. A. Watson, Interview by Ted Anderson, March 14, 2012.

7. V. C. Wynne-Edwards, Low reproductive rates in birds, especially sea-birds, *Acta XI International Ornithological Congress* (1955), p 541.

8. Ibid., p 545.

9. Ibid, p 546.

10. V. C. Wynne-Edwards, The over-fishing principle applied to natural populations and their food-resources: and a theory of natural conservation, *Proceedings of the XII International Ornithological Congress* (1960), p 792.

11. Ibid., p 793.

12. V. C. Wynne-Edwards, The control of population-density through social behaviour: A hypothesis, *Ibis* 101:440 (1959).

13. V. C. Wynne-Edwards, *Animal Dispersion in Relation to Social Behaviour* (New York: Harper, 1962), p 1.

14. Ibid., pp 7–8.

15. Ibid., p 9.

16. Ibid., p 13.

17. Ibid., p 14.

18. Ibid., p 15.

19. Ibid., p 20.

20. H. Buechner, Book review, *Auk* 80:209 (1963).

21. E. M. Nicholson, Special review, *Ibis* 104:571 (1962).

22. C. S. Elton, Self-regulation of animal populations, *Nature* 197:634 (1963).

23. V. C. Wynne-Edwards, "This week's citation classic," Institute for Scientific Information, June 23, 1980.

24. D. Lack, Significance of clutch-size in swifts and grouse. *Nature* 203:98–99 (1964).

25. V. C. Wynne-Edwards, *Nature* 203:99 (1964).

26. D. Lack, *Population Studies of Birds, op. cit.,* pp 311–312.

27. A. Lack, Personal communication, March 13, 2012.

28. A. Watson, *The Ratcatcher* (1997). *The Ratctcher* was an internal newsletter of the staff and students of the Centre of Ecology and Hydrology Banchory Research Station. Banchory, Scotland.

CHAPTER 10

1. D. Lack, *Ecological Adaptations for Breeding in Birds* (London: Methuen, 1968), p 306.

2. D. Lack, *The Life of the Robin* (London: H. F. & G. Witherby, 1943), p 71.

3. B. G. Murray, Jr., Book review, *Auk* 86:778 (1969).

4. Godman-Salvin Medal, *Ibis* 138:606 (1996).

5. D. Lack, EGI Annual Report, *Bird Study* 15:44 (1968).

CHAPTER 11

1. D. Lack, *Ecological Isolation in Birds* (Cambridge, Mass.: Harvard University Press, 1971), p 1.

2. Ibid., p 11.

3. T. R. Anderson, Composition of nestling diets of sparrows, *Passer* spp., within and between habitats, *Acta XVII International Ornithological Congress* 1162–1170 (1980).

4. D. Lack, *loc. cit.*, p 253.

5. R. E. Ricklefs, Book review, *Science* 198:288 (1972).

6. M. L. Cody, Book review, *Auk* 89:907 (1972).

7. P. Medawar, Huxley remembered, *Nature* 254:4 (1975).

8. D. Lack, Trip to California 1935 (typescript), p 34, Lack archive, folder 27.

9. E. Mayr, In appreciation, *Ibis* 115:432 (1973).

10. E. Mayr, Letter to David Lack, March 20, 1946, Lack archive, folder 60.

11. N. Tinbergen, Letter to David Lack, February 26, 1940, Lack archive, folder 155.

12. D. Lack, Obituary of R. E. Moreau, *Ibis* 112:557 (1970).

13. R. E. Moreau, Autobiography, *Ibis* 112:550 (1970).

14. Ibid., p 553.

15. K. Johnson, *The Ibis:* Transformations in a twentieth century British natural history journal, *Journal of the History of Biology* 37:515–555 (2004).

16. D. Lack, *Ecological Isolation in Birds* (Cambridge, Mass.: Harvard University Press, 1971), p 17.

17. P. Ashmole, Interview by Ted Anderson, October 19, 2009.

18. New Fellows of the Royal Society, *Nature* 167:463 (1951).

19. W. H. Thorpe, Appreciation, *Ibis* 115:436 (1973).

20. D. Lack, My life as an amateur ornithologist, *Ibis* 115:423 (1973).

21. D. Lack, Obituary of B. W. Tucker, *Ibis* 93:300 (1951).

22. Ibid., pp 302–303.

23. D. Chitty, *Do Lemmings Commit Suicide? Beautiful Hypotheses and Ugly Facts* (Oxford: Oxford University Press, 1996), p 121. (Name supplied by D. Chitty, Personal communication.)

24. K. Johnson, *loc. cit.*

25. H. R. Witherby, Obituary of A. C. Meinertzhagen, *British Birds* 22:58 (1928).

26. R. Meinertzhagen, Letter to David Lack, February 1, 1945, Lack archive, folder 157.

27. C. D. Darlington, Determined but lonely, *Nature* 220:934 (1969).

28. J. B. S. Haldane, Letter to David Lack, April 25, 1963, Lack archive, folder 223.

29. D. Lack, *Population Studies of Birds* (Oxford: Clarendon Press, 1966), p 300.

30. S. J. Gould and R. C. Lewontin, The spandrels of San Marco and the Panglossian paradigm: a critique of the adaptationist programme, *Proceedings of the Royal Society of London B* 205:581–598 (1979).

31. D. Jenkins and A. Watson, Obituary of V. C. Wynne-Edwards, *Ibis* 119:418 (1997).

32. R. H. MacArthur, Population ecology of some warblers of northeastern coniferous forests. *Ecology* 39:599–619 (1958).

CHAPTER 12

1. Classification based on Paul A. Johnsgard's *Handbook of Waterfowl Behavior* (Ithaca, N.Y.: Cornell University Press, 1965) and doubtless considerably modified by subsequent taxonomic work.

2. D. Lack, *Evolution Illustrated by Waterfowl* (New York: Harper & Row, 1974), p 7.

3. Ibid., p 40.

4. Ibid., p 78.

5. Ibid., pp 81–82.

6. D. Lack, Island biology, *Biotropica* 2:29–31 (1970).

7. D. Lack, *Island Biology Illustrated by the Land Birds of Jamaica* (Berkeley, Calif.: University of California Press, 1976), p 13.

8. A. W. Diamond, Interview by Ted Anderson, March 15, 2010.

9. A. K. Kepler, Interview by Ted Anderson, June 22, 2012.

10. D. Lack, Letter to Tony Diamond, August 11, 1970, Courtesy of A. W. Diamond.

11. A. Lack, Personal communication, June 25, 2012.

12. D. Lack, Christmas letter 1972, Courtesy of C. Lack.

13. Ibid.

14. Ibid.

15. H. Gardner, ed., *The Faber Book of Religious Verse* (London: Faber and Faber, 1972), p 78.

16. W. H. Thorpe, *Ibis* 115:439 (1973).

17. C. Lack, Personal communication, June 30, 2012.

18. Ibid.

19. Peter Lack, Personal communication, June 29, 2012.

20. C. Lack, *op. cit.*

21. C. G. Phillips, David Lack, Sc.D., F.R.S. (obituary), *Trinity College Record, 1972–73*, pp 11–12.

22. E. Mayr, *Ibis* 115:434 (1973).

CHAPTER 13

1. R. H. MacArthur and E. O. Wilson, *The Theory of Island Biogeography* (Princeton, N.J.: Princeton University Press, 1967).

2. D. Lack, The numbers of bird species on islands. *Bird Study* 16:193 (1969).

3. Ibid., p 206.

4. D. Lack, *Island Biology as Illustrated by the Land Birds of Jamaica* (Berkeley, Calif.: University of California Press, 1976), p 71.

5. R. E. Ricklefs, Book review, *Auk* 94:795 (1977).

6. Ibid., p 797.

7. D. Lack, My life as an amateur ornithologist, *Ibis* 115:423 (1973).

8. 2010 Rankings of high schools in the UK based on A-level results, available at www.university-list.net (accessed January 4, 2013).

9. T. Alerstam, David Lack and bird migration—Radar evidence and evolutionary aspects, David Lack Centenary Symposium, Oxford, July 10, 2010.

10. D. Lack, 1946 Report on the Edward Grey Institute of Field Ornithology, p 7, Courtesy of Peter Lack.

11. T. Alerstam, *loc. cit.*

12. New Fellows of the Royal Society, *Nature* 167:463 (1951).

13. D. Lack, My life as an amateur ornithologist, *op. cit.*, p 431.

14. A. Hodgkin, Anniversary address by Sir Alan Hodgkin, P.R.S., *Proceedings of the Royal Society of London B* 183:5 (1973).

15. B. Stonehouse, Interview by Ted Anderson, October 16, 2009.

16. A. Lack, Personal communication.

17. D. Scott, Personal communication, September 28, 2011.

INDEX

Agricultural Research Center (Amani), 36, 39–40, 86, 187, 190

Agricultural Research Council, 86, 87, 134

Alerstam, Thomas, 227, 228

Alexander Library of Ornithology, ix–x, 88, 96, 100–101, 167, 213

Alexander, W. B., 81, 82, 83, 88, 96, 100–101, 214, 227

Alexander Library of Ornithology, ix, x, 88, 96, 100–101, 167, 213

American Museum of Natural History (AMNH), 44, 59, 181, 182

Animal Behaviour Research Group (ABRG), 93, 94, 185, 186

Army Operational Research Group (AORG), 47–49, 60, 62, 84, 189

Ashmole, Myrtle (Goodacre), 109, 134, 167, 174

Ashmole, Philip, 109, 133, 136, 152, 166, 167, 174

Bailey, Roger, 134, 173

Bear Island expedition, 18, 20, 70

Bertram, Colin, 18, 20, 21

Betts, Monica M, x, 87, 88, 97, 101, 133, 134, 161, 162

Breckland Research Unit, 97, 135, 161

British Ecological Society, 49, 62, 170, 195, 201, 230

British Ornithologists' Union (BOU), 3, 136, 147, 148, 149, 166, 167, 168, 170, 171, 185, 188, 190, 197, 198, 201, 203, 229, 231

British Trust for Ornithology (BTO), 52, 73, 82, 83, 86, 94–96, 116, 146, 162, 163, 173, 191, 196, 197, 208, 220

Britten, Benjamin, 8, 9, 119, 215

brood reduction hypothesis, 76, 104, 226

California Academy of Sciences, 45, 57, 58–59, 99, 182, 226

Campbell, Bruce, 13, 118

Cambridge Bird Club, 3, 17, 176, 180, 189, 190–191

Cambridge Five, 10, 32

Charcot, Jean-Baptiste, 21–23

Chitty, Dennis, viii, x, 78, 89, 105, 137, 145–146, 194

Clarke, John, x, 78

clutch size, vii, 51, 62, 64, 76, 92, 138–139, 144, 147, 152, 157, 158, 159, 207, 226

common swift, 72, 76, 86, 93–94, 103, 104–105, 109, 118, 132, 133, 138, 152, 207, 227

competitive exclusion principle, 51, 61, 62, 145, 177, 178, 189, 193, 195

Cornell University, 28–29, 59